SPRINGER SERIES ON ENVIRONMENTAL MANAGEMENT

BRUCE N. ANDERSON

ROBERT W. HOWARTH

LAWRENCE R. WALKER

Series Editors

G.S. Kleppel
M. Richard DeVoe
Mac V. Rawson

Editors

Changing Land Use Patterns in the Coastal Zone

Managing Environmental Quality in Rapidly Developing Regions

With 26 Illustrations

 Springer

Editors:

G.S. Kleppel
Department of Biological
Sciences
University at Albany
State University of New York
1400 Washington Avenue
Albany, NY 12222, USA
gkleppel@albany.edu

M. Richard DeVoe
Sea Grant Consortium
287 Meeting Street
Charleston, SC 29407 USA
rick.devoe@scseagrant.org

Mac V. Rawson
Georgia Sea Grant Program
University of Georgia
Athens, GA 30602-3636 USA
mrawson@uga.edu

Series Editors:

Bruce N. Anderson
Planreal Australasia
Keilor, Victoria 3036
Australia
bnanderson@compuserve.com

Robert W. Howarth
Program in Biogeochemistry
and Environmental Change
Cornell University, Corson
Hall
Ithaca, NY 14853 USA
rwh2@cornell.edu

Lawrence R. Walker
Department of Biological
Sciences
University of Nevada
Las Vegas, NV 89154 USA
walker@unlv.nevada.edu

Cover illustration: Low and moderate density urban development in the headwaters of Boone Hall Creek near Mount Pleasant, South Carolina. Photo by George Steele, South Carolina Department of Natural Resources.

Library of Congress Control Number: 2005931124

ISBN-10: 0-387-28432-X   e-ISBN: 0-387-29023-0
ISBN-13: 978-0387-28432-3

Printed on acid-free paper.

Printed in the United States of America. (Apex/MVY)

9 8 7 6 5 4 3 2 1
springer.com

Jemma Mae

Jemma Mae sells baskets from a stand along the coastal road
just north of Charleston,

Near the very place where she was born,
in a tiny house beside the marsh.

Her mom taught her how to weave the sweetgrass
into baskets

And she points to where two oaks once stood, and
says they made a shady place

Where she could sew the straw.

She says that folks just used to stop and look at baskets
on their way to church or Charleston.

And sometimes someone bought a basket for a sister in Atlanta
or friend in Alabama, or a cousin in New York.

But the road is getting wider now:
all the shade is gone.

Lots of people visit Jemma's stand,
but not from church or Charleston.

Jemma says she's tired and that she don't feel

Much like talking

But then she points a crooked finger
toward the marsh where she grew up.

She talks
about her people

And how their lives are woven
like sweetgrass, into baskets:

Baskets full of stories, baskets full of hope,
and sadness.

Then she points to a construction site
and to where the CVS will be.

And she tells me that her daughter
doesn't want to weave grass baskets.

G.S. Kleppel

June 1999

Dutch Fork, South Carolina

Preface

Each year, numerous edited volumes are produced by scientific publishers that describe or review research in a particular discipline. For the most part, the papers that appear in these volumes represent the efforts of scientists to communicate with other scientists. Many of the authors and most of the readership are faculty, postdocs and graduate students at universities, or researchers at government, academic, or private laboratories. This is certainly an important function of scientific publishing and one to which the present volume will hopefully contribute.

There is, however, another purpose to scientific communication. That purpose, though underrepresented in the professional literature, is crucial to society. In the language of the National Science Foundation, today's science must seek a broader impact. Scientists are becoming increasingly aware of their capacity and responsibility to inform public policy and the decision-making process. An essential role of scientists today is to translate technical data into socially relevant products.

There are few areas of research where the transfer of scientific information to the public sector is more critically needed than in understanding and mitigating the impacts of development on coastal habitats. Each year more than 2 million acres of forests, fields, and farmland are converted to urban, largely suburban, use. On average, the rate of development on the U.S. coastal plain is one-and-a-half times greater than inland. Three of the five fastest growing states in the union are coastal, and ecosystems along our nation's coasts are being lost or modified even before scientists can document their ecological structure and functioning. Despite the use of modern secondary wastewater treatment, retention ponds, and other mitigation techniques, water quality degradation and habitat destruction have become synonymous with what is best described as coastal sprawl.

This book arose from a deep belief among its contributors that science and scientists are important, even crucial, elements in the coastal development paradigm. It is not the mission of scientists to impede coastal development. Rather, it is their job to provide guidance that helps to reduce the impacts of urbanization on critical natural resources, regional cultures, and national coastal heritage. We believe that this is a noble and tractable mission. It is with this motivation that we undertook to produce a volume dealing with land use change in coastal regions. We sought from the beginning to simultaneously address the technical needs of scientists and the practical needs of resource managers, planners, policy-makers, and others who must make decisions that affect people, cultures, the environment, and ecological systems.

In compiling this volume, we sought to provide the reader with the state of knowledge on key issues associated with the responses of coastal systems to rapid urban development. Our goal is to share our science not only with other scientists but also with society-at-large. Contributors include natural and social scientists, empiricists, and modelers. Much of the volume is based on the initial results of a program called the Land Use-Coastal Ecosystem Study (LU-CES) which is supported by the Coastal Ocean Program of the National Oceanic and Atmospheric Administration (NOAA, U.S. Department of Commerce) and administered by the South Carolina Sea Grant Consortium. The scientific mission of LU-CES is to develop protocols and models based on scientific studies of processes that govern the transmission of impacts from changing land use patterns to salt marsh and tidal creek ecosystems in Georgia and South Carolina, with the goal of assisting local decision-makers in their efforts to minimize those impacts while accommodating an increasing coastal population. Because the geographic emphasis of the LU-CES program is on the Southeast, the chapters in the book are regionally focused. The message that emerges, however, is applicable to any rapidly developing coastal landscape in the United States.

There are several individuals whose contribution to the LU-CES program were pivotal. Had they not leant their support to the program, this book would not have been written. First and foremost, we are indebted to U. S. Senator Ernest "Fritz" Hollings and his staff for their support for the nation's coastal and marine resources, and their recognition of the role that science can play in addressing the environmental challenges we face in coastal America. We are also indebted to Dr. James J. Alberts, Ms. Margaret A. Davidson, Esq., Dr. A. Fredrick Holland, Dr. F. John Vernberg, and Dr. Herbert L. Windham for their vision and leadership with the early organization of LU-CES. We appreciate the assistance of our colleagues at NOAA, including Dr. Larry Pugh, Dr. David Johnson, Dr. Kendrick Osgood, Mr. John Wickham, and Dr. Malcolm Meaburn, for their support and guidance. We acknowledge the continuous assistance provided by a panel of resource managers and urban planners who advised the program and helped us to appreciate the important role that environmental research has to play in society. Finally, we thank the many anonymous referees, who reviewed the chapters in this volume.

Changing Land Use Patterns in the Coastal Zone is divided into three parts: I. Trends in Coastal Population Growth—Policies and Predictions; II. Coastal Hydrology and Geochemistry; and III. Contaminants and Their Effects. An important feature of this book is that each chapter following the Introduction (Chapter 1) is preceded by a brief, nontechnical summary. In all but a few cases, these summaries were produced by outreach specialists from the South Carolina Sea Grant Consortium and Georgia Sea Grant College Program, frequently in collaboration with the chapter authors. The intent of the summaries is to translate the technical information and concepts contained in the chapters into text useful to informed readers, including those engaged in coastal development issues. The chapter that follows each summary is traditional in its technical content and focuses on an issue of interest to a particular discipline or group of disciplines. The intended audiences of the technical content of the chapters are scientists, graduate students, resource managers, and other professionals.

In Chapter 1, DeVoe and Kleppel provide a general description of the changes in land use that are taking place nationally and in the southeastern coastal zone. They offer a conceptual framework to guide research aimed at addressing the documented challenges in coastal resource management that rapid land use change, particularly coastal urbanization, creates.

The three chapters in Part I focus on land use policies and their consequences. In Chapter 2, Kleppel, Becker, Allen, and Lu point out that the explosive increase in coastal development that began in the Southeast during the 1970s and continues today as urban sprawl is dictated by federal and local laws that have had serious unintended consequences. The authors go on to describe how modern planning approaches might change development patterns and protect large landscapes and regional cultures. In Chapter 3, Allen and Lu describe the development of models that predict the trajectory of coastal development. They focus on the use of hybrid models that combine several approaches to produce realistic projections of where development is headed. The chapter concludes with a discussion of next generation neural network models that are more accurate and flexible than traditional hybrid approaches. Part I concludes with a conceptual study by Kleppel, Porter, and DeVoe of the potential effects of different urban development styles or typologies on the biodiversity of bottom-dwelling invertebrate communities in the tidal creeks along the rapidly developing coast of South Carolina. The authors argue that traditional urban typologies would require the urbanization of less land in coastal watersheds than suburban typologies, and, therefore, should have a smaller impact on biodiversity than suburban typologies.

In Part II, the authors consider the hydrodynamic and geochemical environments of coastal ecosystems. In Chapter 5, Blanton, Andrade, and Ferriera address the problem of predicting water movement in tidal creeks, which are important both ecologically and economically. Tidal creek hydrodynamics are governed by complex physics. The focus of the chapter is on the development of hypsometric models, which describe the filling and emptying of the creeks during the tidal cycle, from remotely sensed, empirical data. The applications of these models to the identification of locations for future development projects that minimize impacts, and the prediction of future costs (such as dredging) associated with siting decisions show how planners and engineers in coastal communities can use such information to design communities in relatively cost-effective and sustainable ways. In Chapter 6, McKellar and Bratvold discuss the nutrient cycle of coastal, estuarine, and salt marsh systems and their relationship with the watershed. They emphasize the complexity of these systems and discuss the importance of understanding the nutrient budget in order to manage water quality and ecosystem functionality. In chapter 7, Joye, Bronk, Koopmans, and Moore offer an informed perspective on groundwater, which often contributes substantively to the ecology and water quality of coastal systems but which is, at best, poorly studied in many regions. Groundwater transports both nutrients and contaminants to the nation's estuaries, but detailed studies of its dynamics have, for the most part, been lacking. Recommendations are made to resolve the most pressing data needs. In Chapter 8, the conclusion of Part II, Pomeroy and Cai report on oxygen and carbon dioxide dynamics in slat marsh systems and on the use of these variables for predicting anoxia and hypoxia. The chapter contains a complete review of measurement techniques and their applications to resource management.

Part III deals with contaminants of coastal ecosystems and their effects. In Chapter 9, Lee and Maruya catalog the types and sources of toxic contaminants entering coastal and estuarine ecosystems as a result of changing land use patterns in the southeast. Commentary on the constraints (or lack thereof) to the availability and use of pesticides and other contaminants and the challenges that the increasingly impervious landscape will create for contamination prevention and mitigation should provide a wake-up call to coastal communities and the policy-makers who represent them. Chapter 10 by Siewicki discusses key developments in the use of models to predict contaminant-induced ecosystem stress. The directions of current research and needs for future research are examined in detail. Finally, in Chapter 11, Frischer and Verity discuss alternatives to coliform bacteria in the detection of microbial contamination in coastal and estuarine systems. The authors also consider existing and novel microbial and bacteriological approaches to a variety of environmental assessment needs.

We end this volume with an afterword by three recognized leaders in coastal research who have applied scientific data to numerous coastal policy issues. Dr. Geoffrey I. Scott is director of the NOAA Center for Coastal Environmental Health and Biomolecular Research, in Charleston, South Carolina. Dr. A. Fredrick Holland is director of NOAA's Hollings Marine Laboratory, also in Charleston. Dr. Paul A. Sandifer, former director of the South Carolina Department of Natural Resources, is currently senior scientist at the NOAA National Centers for Coastal Ocean Science. Together they propose a new paradigm for coastal research in which coastal policy is driven by science. Objectively gathered scientific data, they suggest, has the potential to resolve conflicts between stakeholders in local development issues. The challenge to the scientific community is to develop data and protocols that respond to society's needs and can be integrated into the decision-making process. The challenge to society is to be willing to accept scientific findings. Can it happen? Some would say it must, if we are to retain the integrity of our nation's coastal heritage in the face of dramatic changes in land use patterns.

January 2006
G.S. Kleppel
M. Richard DeVoe
Mac V. Rawson

Contents

Contributors

Editors
G.S. Kleppel
Biodiversity, Conservation and Policy
    Program
Department of Biological Sciences
State University of New York
    at Albany
Albany, NY, 12222 USA
E-mail: gkleppel@albany.edu

M. Richard DeVoe
South Carolina Sea Grant Consortium
287 Meeting Street
Charleston, SC 29407 USA
E-mail: rick.devoe@scseagrant.org

Mac V. Rawson
Georgia Sea Grant Program
208 Marine Sciences Building
University of Georgia
Athens, GA 30602–3636 USA
E-mail: mrawson@uga.edu

Associate Editors
David A. Bryant
Georgia Sea Grant Program
208 Marine Sciences Building
University of Georgia, Athens GA
E-mail: bryantd@uga.edu

Susan E. Ferris
South Carolina Sea Grant Consortium
287 Meeting Street
Charleston, SC 29407 USA
E-mail: ferris@scseagrant.org

Daniel R. Hitchcock
South Carolina Sea Grant Consortium
287 Meeting Street
Charleston, SC 29401 USA
E-mail: dhitchc@clemson.edu

April L. Turner
South Carolina Sea Grant Consortium
287 Meeting Street
Charleston, SC 29401 USA
E-mail: april.turner@scseagrant.org

Contributors
Jeffery S. Allen
Strom Thurmond Institute of
    Government and Policy Affairs
Silas Pearman Boulevard
Clemson University
Clemson, SC 29634–0125 USA
E-mail: jeff@strom.clemson.edu

Francisco A. L. Andrade
Laboratorio Maritimo da

Guia—FCUL
Estrada do Guincho
2750 Cascais Portugal
E-mail: lmguia@fc.ul.pt

Robert H. Becker
Strom Thurmond Institute of
   Government and Policy Affairs
Silas Pearman Boulevard
Clemson University
Clemson, SC 29634–0125 USA
E-mail: becker@strom.clemson.edu

Jackson O. Blanton
Skidaway Institute of Oceanography
10 Ocean Science Circle
Savannah, GA 31411 USA
E-mail: jack@skio.peachnet.edu

Delma J. Bratvold
Savannah River Ecology Laboratory
University of Georgia
Drawer E., Aiken SC 29808 USA
E-mail: bratvold@musc.edu

Deborah A. Bronk
Virginia Institute of Marine Sciences
College of William and Mary
Gloucester Point, VA 23062 USA
E-mail: bronk@vims.edu

Wei-Jun Cai
Department of Marine Sciences
University of Athens
Athens, GA 30602–3636 USA
E-mail: wcai@uga.edu

M. Richard DeVoe
South Carolina Sea Grant Consortium
287 Meeting Street
Charleston, SC 29407 USA
E-mail: rick.devoe@scseagrant.org

M. Adelaide Ferreira
Laboratorio Maritimo da Guia-IMAR,
   Estrada do Guincho

2750 Cascais Portugal
E-mail: lmguia@fc.ul.pt

Mark E. Frischer
Skidaway Institute of
   Oceanography
10 Ocean Science Circle
Savannah, GA 31411 USA
E-mail: frischer@skio.peachnet.edu

A. Frederick Holland
NOAA, National Center for Coastal
   Ocean Science
Hollings Marine Laboratory
331 Ft. Johnson Road
Charleston, SC 29422 USA
E-mail: fred.holland@noaa.gov

Samantha B. Joye
Department of Marine Sciences
University of Athens
Athens, GA 30602–3636 USA
E-mail: mjoye@arches.uga.edu

G.S. Kleppel
Department of Biological
   Sciences
University of Albany
State University of New York
1400 Washington Avenue
Albany, NY 12222 USA
E-mail: gkleppel@albany.edu

Dirk J. Koopmans
Department of Marine Sciences
University of Georgia
Athens, GA 30602–3636 USA
E-mail: dirk@arches.uga.edu

Richard F. Lee
Skidaway Institute of
   Oceanography
10 Ocean Science Circle
Savannah, GA 31411 USA
E-mail: dick@skio.peachnet.edu

Kang Shou Lu
Strom Thurmond Institute of
    Government and Policy Affairs
Silas Pearman Boulevard
Clemson University
Clemson, SC 29634–0125 USA
E-mail: kshoulu@towson.edu

Keith A. Maruya
Southern California Coastal Water
    Research Project
7171 Fenwick Lane
Westminster, CA 92683 USA
E-mail: keithm@sccwrp.org

Hank N. McKellar, Jr.
South Carolina Department of
    Natural Resources
Freshwater Fisheries Laboratory
1921 VanBoklen Road
Eastover, SC 29044 USA
E-mail: mckellar@dnr.sc.gov

William S. Moore
Department of Geological Sciences
701 Sumter Street
University of South Carolina
Columbia, SC 29208 USA
E-mail: moore@geol.sc.edu

Dwayne E. Porter
Department of Environmental Health
    Sciences
701 Sumter Street
University of South Carolina
Columbia, SC 29208 USA
E-mail: dporter@inlet.geol.sc.edu

Lawrence R. Pomeroy
Institute of Ecology

University of Georgia
Athens, GA 30602–2202 USA
E-mail: lpomeroy@sparrow.uga.edu

Mac V. Rawson
Georgia Sea Grant Program
200 Marine Sciences Building
University of Georgia
Athens, GA 30602–3636 USA
E-mail: mrawson@uga.edu

Paul. A. Sandifer
NOAA, National Center for Coastal
    Ocean Science
Hollings Marine Laboratory
331 Ft. Johnson Road
Charleston, SC 29422 USA
E-mail: paul.sandifer@noaa.gov

Geoffrey I. Scott
NOAA Center for Coastal
    Environmental Health and
    Biomedical Research
219 Ft. Johnson Road
Charleston, SC 29412- 9110 USA
E-mail: geoff.scott@noaa.gov

Thomas C. Siewicki
NOAA Center for Coastal
    Environmental Health and
    Biomedical Research
219 Ft. Johnson Road
Charleston, SC 29412–9110
E-mail: siewicki@noaa.gov

Peter G. Verity
Skidaway Institute of Oceanography
10 Ocean Science Circle
Savannah, GA 31411 USA
E-mail: pete@skio.peachnet.ecu

List of Figures

List of Tables

Introduction—The Effects of Changing Land Use Patterns on Marine Resources: Setting a Research Agenda to Facilitate Management

M. Richard DeVoe and G. S. Kleppel

1.1 Population Growth and Development in the Coastal Zone

1.1.1 National Trends

The coastal United States is an economic engine. Almost 31 percent of the Gross National Product of the United States is generated in coastal counties, and 75 percent of the nation's Gross State Product comes from the coastal states (Colgan 2003). Counties located within 80 kilometers of an ocean or Great Lake, while making up just 13 percent of the landmass of the continental United States, accounted for 50 percent of the nation's population in 2000 and 56 percent of the civilian income in 1999 (Rappaport and Sachs 2001). The near-shore area, which represents 4 percent of the nation's land, produces more than 11 percent of the nation's economic output (Colgan 2003).

Economic growth in the nation's coastal regions is pushing the demand for land conversion from open space, farmland, and wildlife habitat to residential and commercial uses (Colgan 2003). More than one-fourth of all land conversions from rural to urban and suburban uses over the last three centuries has occurred in the past 15 years, and between 1982 and 1992, land was developed at 1.8 times the rate of population growth; from 1992 to 1997 that factor had grown to 2.5 (Beach 2002).

In 2003, more than 153 million people—about 53 percent of the population of the United States—lived in the coastal regions of the United States (NOAA 2004), and by 2015, the population of these regions will reach 165 million (W&PE 2003, in NOAA 2004). Coastal counties average 300 people per square mile (excluding Alaska), an increase of 65 people per square mile since 1980 and more than three times the national average of 98 people per square mile (NOAA 2004). Population density is expected to increase by another 13 people per square mile (or 4 percent) by 2008 (NOAA 2004).

During the last decade, 17 of the 20 fastest growing counties in the United States were located along the coast, and this region accounts for 23 of the 25 most densely populated counties in the country (NOAA 2004). Because the coast will retain its current population share of growth, it will absorb more than half of the nation's population growth in the next several decades (Beach 2002).

Development is placing tremendous pressure on the natural resources of the coastal United States—the beaches, waterways, oceans, rivers, and estuaries where people come to live, work, and play. The more people, the more infrastructure; the more infrastructure, the more impervious surfaces. Impervious surfaces include paved areas, such as roads and parking lots, as well as roofs that rain cannot penetrate. Increases in impervious cover associated with development increase both the volume and the rate of storm water runoff. When the amount of impervious surface in a watershed area reaches 10 percent or more, the resources within that watershed become altered and sometimes impaired (Beach 2002 and references therein).

Poorly managed growth can also translate into increased taxes and spending for urban infrastructure and services (O'Hara 1997). Beach (2002) described reforms needed to address the impacts and costs of growth on coastal and ocean resources, and ideas about how to apply some of those reforms to prevent or mitigate sprawl are provided by Kleppel, Becker, and Allen (manuscript) in this volume.

1.1.2 The Southeastern U.S. Coastal Region

From the interior basins to the coastal margins, natural processes and human activities in the southeastern United States are affected by water flow and its role in determining the transport and fate of materials and the structure of ecosystems. Inputs of freshwater from rivers, ground-water, and rainfall vary spatially and temporally. Associated with the volumes of water delivered to the coastal ocean are variable loads of sediment, nutrients, and pollutants. The inputs of freshwater and materials interact with the coastal ocean to influence processes such as local circulation patterns, sediment accumulation and transport, and habitat quality and stability for marine and estuarine species.

The South Atlantic Bight along the southeastern United States covers an area of more than 90,600 square kilometers, and the continental coastline that borders it extends for almost 2,000 kilometers along Florida, Georgia, and the Carolinas. Diverse and extensive estuarine systems in the region cover an estimated 11,600 km^2 (U.S. EPA 2002). Some 47,000 km of tidal shoreline rim the more than 300 estuaries in the southeastern and Gulf coasts of the United States, which supported about $850 million in commercial fishery landings in 2003 (pers. comm. NOAA National Marine Fisheries Service, Fisheries Statistics Division) and $3.4 billion in recreational fisheries annually (NOAA 1990). At least 96 percent of the commercial and 70 percent of recreationally important fish and shellfish in the southeastern region require estuaries and near-shore marine habitats during their life cycles. This region can be considered a large marine ecosystem with common physical and demographic attributes, and multi-disciplinary research efforts to examine the relationships between land use and coastal ecosystem integrity on a variety of scales are needed.

The coastal counties of the southeastern United States have seen unprecedented growth over the last 30 years. Population in southeastern coastal counties increased by 64 percent between 1970 and 1990 (U.S. Bureau of the Census 1996, in U.S. EPA 2002). The Southeast region exhibited the largest rate of population growth in the country (58 percent) between 1980 and 2003; three of the ten states with the highest percent change in population growth in the United States were located in the Southeast (NOAA 2004).

It is expected that during the next 20 to 30 years, the Southeast will continue to experience high population growth (DeVoe and Kleppel 1995; NOAA 1999, 2004). The majority of this growth will occur along the coast and will be associated with an in-migration of both retirees and job seekers (U.S. Census Bureau 1998; NOAA 2004). Indeed, the U.S. Census Bureau (1998) projected that 11 million additional people, 14 percent of the post-World War II baby boom generation, will reside in the Carolinas and Georgia by 2025. While silvaculture is expected to remain the principal industry in the Southeast during this period, retirement and tourist-oriented businesses will dramatically increase the economic investment and level of development within the region.

Significant impacts to the landscape, estuarine water quality, and coastal ecosystem integrity are predicted as a result of urbanization

(Kleppel and DeVoe 2000; Beach 2002). Growth and development are already placing enormous pressure on coastal resources, watersheds, and the adjacent coastal ocean. The impacts of rapid, often loosely managed growth can drastically alter the cultures and quality of life of people in the Southeast; the importance of understanding and addressing these challenges should not be underestimated.

Perhaps the most tangible sign of urban impacts is the closure of shellfish beds due to contamination, since this indicates not only where human activities have degraded environmental quality, but also where watershed management has failed (Fletcher et al. 1998). More than 30 percent of the shellfish harvesting waters of Georgia, South Carolina, and North Carolina have been closed due to fecal coliform contamination. Similarly, severe restrictions on fish harvest can indicate that living marine resource management has had limited success. Fish advisories due to chemical contamination also serve as indicators of ineffective environmental management. These examples illustrate the clear need to examine the relationship between land use and the integrity of coastal ecosystems.

The opportunity for managed growth and scientifically sound decision-making still exists in the southeastern United States. Unfortunately, much of the scientific knowledge needed to inform the decision-making process either does not exist, or has yet to be delivered to decision-makers in a usable form. The effects of watershed alteration on coastal resources in the Southeast are, in general, poorly understood and inaccessible to watershed and coastal resource managers. New management strategies and techniques will be needed by urban and regional planners and resource managers as they seek to address the economic, public health, environmental quality, and quality of life issues emerging as a result of rapid population growth and development.

1.2 Scientific and Management Issues

1.2.1 Pertinent Resource Management Challenges

Resource managers recognize that coastal ecosystem integrity could be compromised by the high rate of population growth and associated development predicted for the Southeast over the next two decades. They have identified the following challenges as critical to the management of coastal and marine resources:

1. **Distribution of population.** Much of the growth and development that will take place in the coastal Southeast during the next two decades will occur in environmentally sensitive areas. There is little policy to guide the increasingly strained relationship between resource protection and economic development.
2. **Urbanization.** Presently, more than 60 percent of the land in coastal watersheds in the region is covered with natural vegetation or used for silvaculture or agriculture. A significant portion of this landscape could be converted to urban use during the coming decades. As a result, impervious surface area will increase and the water budget and contaminant transport mechanisms will change. The consequences of such changes in land use patterns for coastal ecosystems are poorly understood.
3. **Surface and groundwater resources.** The management of water to ensure adequate quantity and quality will drive resource decision-making for several decades. The development of management strategies based on the accurate prediction of impacts and outcomes is crucial to the maintenance of water quality and coastal ecosystem integrity. Little is known about the distribution and chemical composition of groundwater resources in many parts of the region, nor is there a clear understanding of the physical, chemical, and biotic mechanisms by which contaminants, nutrients, and other constituents are transported to and processed within coastal ecosystems.
4. **Land use decision-making.** Striking a balance between increased population growth and development and conservation of natural resources requires that those responsible for managing growth work with those managing natural resources. However, while much of the authority for natural resource management rests with officials at the federal and state levels, most land use decisions are made at the county or municipal level (Dale et al. 2000). This "disconnect" constrains land use planning, which is further hampered by inadequate exchange of technical and scientific information.
5. **Translation of science-based information.** Data and information from scientific research must be packaged and delivered in formats that meet the specific needs of the management and planning communities. That translation must be rapid, as changes in land use

and concomitant effects on environmental quality and ecosystem and cultural integrity are occurring at high rates. The current rate of conversion of scientific results into management applications is many times too slow to be of value in attempts to balance growth with the protection of socioeconomic and environmental values.

6. **Education and outreach.** Most growth and development decisions are influenced by public constituencies. A clear understanding of the costs and benefits of growth is lacking throughout the public sector. Methodologies for collecting, synthesizing, and disseminating scientific data to inform land use decisions must be improved. Environmentally and economically enlightened decision-making is crucial to sustainable development and resource conservation in the region.

1.2.2 Matching Management Needs with Science-Based Information

Any effort to understand the relationships between land use and the condition of marine ecosystems must be multi-disciplinary, involving both the natural and social sciences. In developing such a program, the challenges to the scientific community include:

1. **Varying time and space scales.** Temporal and spatial variability scales determine the outcomes of numerous biotic, chemical, and physical processes. Identifying the factors that drive these variability scales as they relate to the transport, fate, and effects of land-derived contaminants in estuarine and coastal waters is among the most important challenges in environmental science and management.
2. **Natural variability.** The behavior of any ecosystem is driven by variable physical, chemical, and biological processes. The challenge is to document and model these natural variations to distinguish them from anthropogenic variability.
3. **Multiple land use activities.** In order to understand the relationships between land use patterns and ecosystem integrity, "signals" or indicators of specific land uses or cover classes must be identified.
4. **Nature of scientific investigation.** Studies of coastal ecosystem responses to changing land use patterns must engage a variety of disciplines. The challenge is to ensure that integration across disciplines and between research and outreach is achieved. A strong project management structure can prevent researchers from becoming

advocates of their own work plans rather than of the project as a whole (Auyong 1998).

5. **Science to management.** The scientific community faces enormous challenges when it comes to designing and implementing research programs to answer questions posed by resource managers and coastal decision-makers. The science-management "disconnect" manifests itself in several ways:

   a. **Time frame.** Decisions by resource managers and coastal decision-makers (e.g., planners, local government officials) tend to be time sensitive and time limited. Often decisions are made without adequate data on hand. Conversely, scientists often require much more time than is available to the decision-maker. Furthermore, scientific data may not be available in the synthesized form needed by the decision-maker. It has been difficult to incorporate ecological values in land use decision-making because science and management are constrained by different time and spatial scales (Hulse and Ribe 2000).

   b. **Communication.** While decision-makers "manage" people, scientists "manage" research. The challenge for the decision-maker is to clearly articulate management needs for the scientist and, for the scientist, to understand the questions, translate them into appropriate research strategies, and ultimately communicate the results back to the decision-maker in a timely fashion and in a usable form.

   c. **Product delivery to decision-makers.** Although the specific products of any research program become more clearly defined as it matures, several general directions for product development can be foreseen. Many solutions take the form of conceptual models that provide land use and development scenarios to guide planning and decision-making. In addition, several approaches, such as GIS, can be used to gather, organize, and deliver data to managers and decision-makers.

1.3 Linking Land Use to Ecosystem Health

1.3.1 A Conceptual Model

Variability in the types and amounts of inputs from terrestrial sources is often due to changing land use practices and water management strategies. In the Southeast, water inputs have varied because of impoundment and increasing use of aquifers for agriculture, landscape irrigation, and other human uses. Future demands for water may result in major interbasin transfers that could shift large quantities of water from the South Atlantic to the Gulf of Mexico. Landscape alterations have led to increased riverborne sediment loading from accelerated soil erosion; for example, coastal turbidity is largely attributable to deforestation and intensive agriculture, much of which began in the eighteenth and nineteenth centuries. The influences of watershed alterations on coastal ecosystems are poorly understood. Further, there is little information available for developing large-scale management strategies for watersheds or coastal environments. A novel approach to planning and management is needed to balance the impacts of rapid growth with resource conservation.

To predict how changing land use patterns and anthropogenic inputs from coastal population growth will affect habitat quality and living marine resources, a functional understanding of the biogeochemical processes common to the entire system as well as those that are unique to individual ecosystems is needed.

1.3.2 System Compartments

As a general conceptual model, coastal regions in the southeastern United States can be divided into distinct "compartments":

- The inner shelf region of the coastal ocean
- Riverine estuaries
- Nonriverine estuaries
- Salt marsh-tidal creek complexes
- Rivers and their associated watersheds and landscapes

Defining biogeochemical processes within each compartment of the coastal system and the particulate and chemical fluxes within and among compartments can provide information useful to the management of aquatic and coastal resources. Characterization of these fluxes into, within, and through compartments is critical for investigating both natural and anthropogenic impacts. As materials are transported from land to sea, they transit a number of compartments from watersheds and rivers, through salt marshes and tidal creeks, into estuaries, and ultimately the coastal ocean. As salinity gradients are crossed, changes occur in transport mechanisms and in the forms the materials take. Understanding these fluxes and the processes that drive them is critical to predicting and managing the loading parameters and carrying capacities of contaminants in coastal ecosystems.

1.3.3 Compartmental Interfaces

Interfaces are boundaries between compartments where sedimentary and transport processes change. For example, alteration in the direction of transport occurs along small tidal creeks due to the distortion of the diurnal internal tidal wave (Pietrafesa et al. 1985; Blanton and Gross 1998). The character and strength of wind-generated and density-driven currents also change at the interfaces between compartments (Blanton et al. 1989a).

Tidal creek-salt marsh complexes are the primary linkages between upland (including freshwater) and estuarine environments. Tidal creeks function as repositories for sediments, organic materials, and chemical contaminants. Urbanized tidal creek watersheds tend to have large amounts of impervious surfaces in them. The tidal creeks that drain these watersheds exhibit altered hydrology (as indicated by salinity range), increased levels of chemical contaminants, altered growth or reduced fecundity of marine benthic organisms, and frequent and sometimes severe hypoxic events (Cai and Pomeroy 1998; Lerberg et al. 2000; Holland et al. 2004).

It is impractical to attempt to monitor all coastal systems from their headwaters to the estuaries on a routine basis. It is necessary, therefore, to identify important systemic interfaces on appropriate scales through which the status of compartments can be evaluated. By focusing on these "compartmental interfaces" of the system, the effects of various land use practices and the influences of changing land use patterns on the physical, chemical, and biological attributes of the system can be identified.

If we assume that land use activities alter stream flows, hydrodynamics, and the loading of both natural and man-made materials to these ecosystems, and that their resultant impacts on biotic resources and processes will eventually be reflected in the structure and functioning of ecosystems, two unifying themes emerge. The first is that different types and intensities of development will result in different impacts to ecosystems and living marine resources. The second is that different types of watersheds and coastal systems will respond differently to development pressures. It is therefore important to define: (1) compartmental interfaces where management actions can intervene to mitigate impact; (2) indicators representative of ecosystem condition; and (3) the dynamic interactions within and across compartments, expressed at the interfaces. The intent should not be to study every system, but to conduct sufficient regional-scale research to develop conceptual models that can be extrapolated to a range of system types within the region.

Efforts to understand the region and the changes taking place within it as a function of human population growth should involve both retrospective analyses (see, for example, Kleppel and DeVoe 1999) and new research. Such efforts also require a strategy that recognizes both the longitudinal breadth of the region (from watersheds to the coastal ocean) on scales at which various components of the system function, and the multi-dimensional nature of the system. Given the size, complexity, and multiplicity of scales that characterize coastal ecosystems, it becomes clear that most of the data acquired from the system will be spatially and temporally discontinuous. It is impossible to sample the entire system continuously or synoptically (Cowen et al. 1998). Instead, the approach should seek to identify the "control points" or convergences within the system where the greatest amount of information regarding anthropogenic impacts can be obtained. This is the idea behind the compartmental interface approach (DeVoe and Kleppel 1995; Kleppel and DeVoe 1996; see examples in Table 1.1).

Compartmental interfaces represent hydrographic junctions at which changes in material fluxes may represent an integrated system response to anthropogenic activities landward of the interface, and are locations that can be used to index system health and condition. Processes or variables assessed at interfaces represent the response of the system integrated over important spatial or temporal scales (Wiegert 1986; Kneib 1994). Effective resource management depends on how well the relationships between process and scale are understood.

1.3.4 Functional Components

For any compartmental interface, a progression or flow can be established that defines the functional aspects of some parameter at that interface. Land use-coastal ecosystem studies should identify markers representative of classes of materials that affect coastal ecosystems for

Table 1.1. Examples of Compartmental Interfaces.

Compartmental Interface	Pertinent Scales	Variables or Processes	Examples of Representative
Extent of Tidal Intrusion	Days to Months	10s to 100s Km	Riverborne particles runoff; Drainage basin scale inputs
Upland/Marsh/ Creek Boundary	Hours to Seasons	Meters	Mixtures of land-based dissolved and particulate materials; freshwater inflow
Tidal Creek/ Riverine Boundary	Hours to Seasons	<1 to10s Km	Fisheries recruitment; Exposure to contaminants
Turbidity Maxima	Hours to Days	1 to 10 Km	Mixtures of dissolved and particulate materials
Estuarine Inlets	Hours to Months	<Km	Fisheries recruitment; Estuarine/ coastal exchanges
Coastal Boundary Zone	Hours to Years	1 to 10s Km	Plume dynamics; Sediment transport; Alongshore transport

the purpose of developing and validating quantitative and predictive models. The influence of a specific contaminant will depend upon the nature of its loading, transport, and ultimate fate. Further, the degree of exposure to this contaminant by living organisms will depend on a number of biogeochemical processes, and hence the effect of that material will be determined by the progression of events.

A cornerstone of an integrated research program, and a mechanism around which a diverse array of research activities can be organized, is an effort to conduct a uniform characterization of the functional components of the coastal marine system(s) selected for study. This necessitates research that measures anthropogenic inputs that are quantitatively linked to land use.

1.4 Studying the Land-Water Interface

1.4.1 Scaling Factors

A component of this conceptual model is the focus on ecological studies on the scales at which most land use decisions are made, and where materials that originate from land use activities first enter the marine ecosystem. In the southeastern United States, these are salt marshes and tidal creeks.

Scaling factors influence the relationships between land use patterns, ecosystem function, and the transport, fate, and effects of contaminants within these systems. One must be aware of the time and space scales across which the compartmental interface paradigm is framed and the spatial dimensions of the landmass (i.e., the watershed or subwatershed) that impacts a particular salt marsh-tidal creek system.

With a mechanistic understanding of the processes that govern the transport, fate, and effects of specific "land use indicators," the feasibility of modeling impacts to coastal ecosystems increases. The processes that must be understood to obtain such a mechanistic perspective include, but are not limited to, those that govern associations between contaminants (metals, toxic organics) and substrates (e.g., humic acids), estuarine metabolism, and nutrient dynamics. Examples of such indicators of land use impacts in salt marshes and tidal creeks the southeastern United States are provided in Table 1.2.

Table 1.2. Possible Land Use Indicators.

Indicator	Application	Reference
O_2/CO_2	Estuary metabolism	Cai and Pomeroy 1998
Dissolved DIN, DON, DIC, DOC, CH_4, CO_2, N_2O in aquifer	Potential for eutrophication; discriminate surface from ground water inputs; possibly discriminate agricultural from domestic nutrient sources	Joye et al. 1998

(*Continued*)

Table 1.2. Possible Land use Indicators. (Continued)

Indicator	Application	Reference
Uranium, ^{210}Pb, ^{137}Cs, dissolved Si	Cultural eutrophication	Windom et al. 1998
E. coli analyses: pulsed gel electrophoresis, ribotyping, fatty acid composition, multiple antibiotic resistance	Distinguish human from other animal wastes; distinguish domestic, farm and wild animal fecal coliform	Fletcher et al. 1998
Specific bacteriophages	Certain bacteriophages attack bacteria specific to human sources (e.g., sorbitol-fermenting bifido-bacteria)	Fletcher et al. 1998
Specific microbial taxa	Urban runoff, eutrophication, chemical contaminants	Fletcher et al. 1998
Luminescent bacteria	Toxic contaminants	Fletcher et al. 1998
Phytoplankton composition	Nutrient distributions	Fletcher et al. 1998
Veterinary pharmaceuticals	Agriculture	Shaw and Chandler 1998
Metal phase; organic C/fulvic and humic acids	Discriminate forested, agricultural and urban inputs	Shaw and Chandler 1998
PAHs	Particular forms associated with urbanization	Kucklick et al. 1997
Specific pesticides	Differentiate agriculture from suburban land use	G. Scott pers. comm.

1.4.2 Process Studies

There is considerable information about the concentrations and distributions of nutrients and contaminants in the estuaries and watersheds of the southeastern United States. However, the processes that regulate these distributions or their impacts in coastal ecosystems are not well understood. Inasmuch as most contaminants in coastal ecosystems occur as a function of particular activities on land, the relationship between land use and contamination is also poorly understood. The processes and mechanisms that govern the associations between changing land use patterns and the integrity of coastal ecosystems require additional characterization and evaluation for their broader predictive power.

1.4.3 Modeling

Coastal ecosystems along the southeastern United States are influenced by one of the most energetic tidal regimes on the eastern seaboard. Tidal and buoyancy-driven processes dominate the circulation patterns of the region's estuaries. However, small, poorly understood asymmetries in transport actually determine the net fluxes of materials. Models that seem capable of resolving local circulation patterns within salt marsh-tidal creek systems are accessible without major investment in software development. Information on the contribution and influence of groundwater on local circulation is a critically important, but largely missing element in circulation and transport models and in land use-ecosystem studies in general (Joye et al., this volume).

Whether they are newly created or generated through modification and adaptation of existing models, the development and evaluation of both conceptual and process models to better correlate linkages among development activities, material flux, habitat quality, and the condition of living marine resources is crucial to the successful study of land use change dynamics. Attempts to calibrate these models will identify shortcomings in both the quality and coverage of the existing data and the functionality of the models. It is unlikely that existing models will be adequate to describe all relevant processes in southeastern marine and coastal systems or address all pertinent management issues. Therefore, linked or hybrid modeling approaches should be considered.

1.5 Conclusion

The rapid rate of land use change along the coasts of the United States, largely in the form of urbanization, requires a new approach to environmental and ecological research. That approach must respond to the needs of resource managers and decision-makers for information about the impacts of coastal development in the context and on the time scales appropriate to local development pressure. The information derived from research must be delivered in a form that is understandable to the nonscientist. This challenge requires an approach to ecological and environmental research that is targeted and interdisciplinary. It requires that the research be followed by the rapid translation of results into scenario-based models understandable by managers, planners, and policy makers.

The conceptual framework around which such an approach is built involves the identification of interfaces or boundaries between compartments in transects from the watershed to the coastal ocean. In the southeastern United States, for example, the salt marsh-tidal creek complex represents one such key interface. Researchers must identify markers or "indicators" of specific land use attributes, such as particular kinds of chemical contaminants, and document the transport, fate, and effects of these indicators across the interfaces. Ultimately, the data must be compiled in a format from which alternative land use scenarios can be modeled and evaluated.

Creating a conceptual framework within which to understand the interactions between land use and coastal ecosystems will require an improved understanding of how coastal ecosystems process anthropogenic materials. This in turn will require an improved understanding of the spatial and temporal constraints on loading, which implies an improved knowledge of the interaction between surface and groundwater.

In the spirit of what is proposed here, the following chapters summarize the accumulated knowledge on regional coastal ecosystem attributes associated with the salt marsh-tidal creek complexes of the southeastern United States. Gaps in that knowledge, where additional research is needed to address issues associated with coastal development, are also identified. Finally, each chapter is preceded by a summary written without the jargon and technical detail that often limits the usefulness of scientific results by nonscientists, in an effort to provide the timely transition of fundamental scientific information to the public.

Acknowledgments

The authors wish to acknowledge the time, energy, and assistance of James J. Alberts, Margaret A. Davidson, A. Frederick Holland, Mac V. Rawson Jr., Geoffrey I. Scott, F. John Vernberg, and Herbert L. Windham, all of whom provided valuable insight and assistance with the development, conceptualization, and implementation of the ideas expressed in this paper, and to the NOAA Coastal Ocean Program for providing support under grant NA960P0113.

References

Auyong, J. 1998. Multidisciplinary projects—challenges and opportunities. *The Coastal Society Bulletin* 20(2):16–19.

Beach, D. 2002. Coastal sprawl—the effects of urban design on aquatic ecosystems in the United States. Pew Oceans Commission, Arlington, VA. 31 p.

Blanton, J. O. and T. F. Gross. 1998. LU-CES State-of-knowledge report on physical oceanography. Land Use-Coastal Ecosystem Study, Charleston, SC. 42p. http://www.lu-ces.org/documents/ SOKreports/physicaloceanography.pdf.

Blanton, J. O., L.-Y Oey, J. Amft and T. N. Lee. 1989a. Advection of momentum and buoyancy in a coastal frontal zone. *Journal of Physical Oceanography* 19:98–115.

Cai, W.-J. and L. Pomeroy. 1998. State of knowledge of respiratory processes and net productivity of intertidal marshes of South Carolina and Georgia. Land Use—Coastal Ecosystem Study, Charleston, SC. 23p. http://www.lu-ces.org/documents/SOKreports/respiratoryproc_ netproduct.pdf.

Colgan, C. S. 2003. The changing ocean and coastal economy of the United States: A briefing paper for conference participants. Presented at Waves of Change: Examining the Roles of States in Emerging Ocean Policy, a National Governors Association Center for Best Practices Conference, September 3. 18 p.

Cowen, D. J., J. R. Jensen, and M. Hodgson. 1998. State of knowledge on GIS databases and land use/coverage patterns: South Carolina. Land Use—Coastal Ecosystem Study, Charleston, SC. 97p. http:// www.lu-ces.org/documents/SOKreports/landuse_landcover.pdf.

Dale, V. H., S. Brown, R. A. Haeuber, N. T. Hobbs, N. Huntly, R. J. Naiman, W. E. Riebsame, M. G. Turner, and T. J. Valone. 2000. Ecological principles and guidelines for managing the use of land. *Ecological Applications* 10(3):639–670.

DeVoe, M. R. and G. S. Kleppel. 1995. The South Atlantic Bight Land Use-Coastal Ecosystem Study (LU-CES): Conceptual framework. South Carolina Sea Grant Consortium, Charleston, SC. (Unpublished manuscript; 28p.)

Fletcher, M., P. G. Verity, M. E. Frischer, K. A. Maruya, and G. I. Scott. 1998. Microbial indicators of phytoplankton and bacterial communities as evidence of contamination caused by changing land use patterns. Land Use-Coastal Ecosystem Study, Charleston, SC. 100p. http://www.lu-ces.org/documents/SOKreports/microbialindicators.pdf.

Holland, A. F., D. M. Sanger, C. P. Gawle, S. B. Lerberg, M. S. Santiago, G. H. M. Riekerk, L. E. Zimmerman, and G. I. Scott. 2004. Linkages between tidal creek ecosystems and the landscape and demographic attributes of their watersheds. *Journal of Experimental Marine Biology and Ecology* 298:151–178.

Hulse, D. and R. Ribe. 2000. Land conversion and the production of wealth. *Ecological Applications* 10(3):679–682.

Joye, S. B., D. A. Bronk, D. J. Koopmans, and W., S. Moore. (This volume.) Evaluating the potential importance of groundwater-derived carbon, nitrogen, and phosphorus inputs to South Carolina and Georgia coastal ecosystems. In, Kleppel, G. S., M. R. DeVoe, and M. V. Rawson (eds.), Changing Land Use Patterns in the Coastal Zone. Springer-Verlag, New York, NY.

Kleppel, G. S. and M. R. DeVoe. 1996. The South Atlantic Bight Land Use-Coastal Ecosystem Study (LU-CES): Report of a planning workshop. South Carolina Sea Grant Consortium, Charleston, SC; University of Georgia Sea Grant College Program, Athens, GA; and NOAA Coastal Ocean Program, Washington, DC, Savannah, GA, June 23–24, 1996. 11 p. http://www.lu-ces.org/documents/Minutes/plnwkshprpt.pdf.

Kleppel, G. S. and M. R. DeVoe. 1999. The South Atlantic Bight Land Use-Coastal Ecosystem Study (LU-CES): The state of knowledge on issues pertinent to the program mission—a synthesis. 21p. http://www.lu-ces.org/documents/SOKreports/sokrptsynthesis.pdf.

Kleppel, G. S. and M. R. DeVoe. 2000. A sense of place: The people factor. *South Carolina Wildlife* 47(1):32–42.

Kneib, R. T. 1994. Spatial pattern, spatial scale, and feeding in fishes, pp. 171–185. In, Stouder, D. J., K. L. Fresh, and R. J. Feller (eds.), Theory and Application in Fish Feeding Ecology. Belle W. Baruch Library in Marine Science 18, University of South Carolina Press, Columbia.

Kucklick, J. R. et al. 1997. Factors influencing polycyclic aromatic hydrocarbon distributions in South Carolina estuarine sediments. *Journal of Experimental Marine Biology and Ecology* 213:13–29.

Lerberg, S. B., A. F. Holland, and D. M. Sanger. 2000. Responses of tidal creek macrobenthic communities to the effects of watershed development. *Estuaries* 23(6):838–853.

National Oceanic and Atmospheric Administration (NOAA). 1990. Earth is a marine habitat: Habitat conservation program. U.S. NOAA, National Marine Fisheries Service, Washington, DC. 8 p.

National Oceanic and Atmospheric Administration (NOAA). 1999. Trends in U.S. coastal regions, 1970–1998. Addendum to the proceedings, trends and future challenges for U.S. national ocean and coastal policy. NOAA National Ocean Service Special Projects Office, Silver Spring, MD. 31 p.

National Oceanic and Atmospheric Administration (NOAA). 2004. Population trends along the coastal United States, 1980–2008, by Crossett, K. N., T. J. Culliton, P. C. Wiley, and T. R. Goodspeed. NOAA National Ocean Service Special Projects Office, Silver Spring, MD. 47 p.

O'Hara, F. 1997. The cost of sprawl. Marine State Planning Office, Executive Department, Bangor, ME. 20 p.

Pietrafesa, L. J., J. O. Blanton, J. D. Wang, V. Kourafalou, T. N. Lee, and K. A. Bush. 1985. The tidal regime in the South Atlantic Bight, pp. 63–76. In, Atkinson, L. P., D. W. Menzel, and K. A. Bush (eds.), Oceanography of the Southeastern United States Continental Shelf (Coastal Estuarine Sciences 2). American Geophysical Union, Washington, DC.

Rappaport, J. and J. D. Sachs. 2001. The U.S. as a coastal nation. Research Division, Federal Reserve Bank of Kansas City. 35 p. plus figures. http://www.kc.frb.org/publicat/neswkpap/rwp01–11.htm.

Shaw, T. J. and G. T. Chandler. 1998. Biogeochemical processes and toxicant impacts from land uses affecting the "head of tide." In, South Atlantic Bight Land Use—Coastal Ecosystem Study, Charleston, SC. 27p. http://www.lu-ces.org/documents/SOKreports/biogeochemical_processes.pdf.

U.S. Census Bureau. 1998. Population projections, 1995–2025. Current population reports. U.S. Department of Commerce, Washington, DC. 6p.

U.S. Census Bureau. 2000. Population projections for South Carolina, p. 26. In Statistical Abstract of the United States, Washington, DC. 62p.

U.S. Environmental Protection Agency (U.S. EPA). 2002. Southeast coastal condition, pp. 88–102. In, National Coastal Condition Report, EPA-620/R-01/005. Washington, DC.

Weigert, R. G. 1986. Modeling spatial and temporal variability in a salt marsh: Sensitivity to rates of primary production, tidal migration and microbial degradation, pp. 405–426. In, Estuarine Variability, Proceedings of the Eighth Biennial International Estuarine Research Conference. Academic Press, Orlando, FL.

Windom, H. L., H. N. McKellar Jr., C. R. Alexander Jr., C. A. Abusam, and M. Alford. 1998. Indicators of trends toward coastal eutrophication. Land Use–Coastal Ecosystem Study, Charleston, SC. 56p. http://www.lu-ces.org/documents/SOKreports/coastaleutrophication.pdf.

Woods and Poole Economics, Inc. (W&PE). 2003. 2003 Desktop Data Files. W&PE, Washington, DC.

Part I

Trends in Coastal Population
Growth—Policies and Predictions

2

Trends in Land Use Policy and Development in the Coastal Southeast

G.S. Kleppel, Robert H. Becker, Jeffery S. Allen, and Kang Shou Lu

Summary by April L. Turner

With an estimated 50 to 75 percent of the growth and development in the southeastern United States during the next 30 years projected to occur along the coastal plain, traditional southern communities are being replaced by retirement communities and exclusive vacation resorts. Natural resource managers, urban and regional planners, and social scientists are concerned that the influx of new residents and visitors to this newly gentrified coastal zone will result in the loss of cultural identity and environmental quality. While the environmental decline and cultural changes associated with urban (particularly sub-urban) development are certainly the results of the physical disruption of the landscape and the cultures thereon, ultimately these disruptions are created by the policies that govern development. If we examine the history of development in the Southeast (or nearly anywhere, for that matter) we find that federal, state, and local policies that regulate and shape land use practices have had consequences that were largely unintended by those who framed the policies. There are few more compelling examples of this than the National Flood Insurance Program (NFIP) of 1968, or the emergence of local single-purpose zoning ordinances created by the state Standard Zoning Enabling Act (SZEA) of 1926.

Following the Civil War, the Southeast experienced one of the lowest regional economic development rates in the nation. This facilitated the concentration of large parcels of land in the hands of a few very wealthy individuals and corporations, largely for their private use.

One outcome of this privatization of coastal property was a reduction in coastal population.

Over the past three decades, population growth on the coastal plain of South Carolina has exploded, and coastal destinations in the Carolinas and Georgia are experiencing an enormous increase in tourism and in retirement and vacation-home development. Given the limited availability of developable land on the coast, growth is tracking inland along rivers, lakes, and major highways.

To predict the consequences of development in the coastal zone, one must also understand the policies that drive it and the implications of those policies for human behavior. Historically, siting decisions for coastal development were governed by a consideration of risk. Abatement of risk (by building in places protected from storms and floods) was a mitigating factor in early land use decisions. However, the NFIP of 1968, which underwrote flood insurance for all borrowers with federally backed mortgages, transferred these risks from developers and lenders to taxpayers. Thus, the NFIP and subsequent federally funded flood protection opened up the coastal real estate market to development by transferring risk to the public.

At the level of local government, zoning and subdivision regulations have defined the layout of the built environment since the end of the Second World War. The concept of single-purpose zoning was introduced as part of the SZEA of 1926. Subsequent ordinances for parcel subdivision and infrastructure development have been significant factors in determining land use patterns in coastal South Carolina. Single-purpose zoning, based on the concept of "consistent use," coupled with pressure from special interests (e.g., oil companies), has led to increased reliance on the automobile for transportation, and severe constraints on pedestrian and public transit opportunities. Such policies have isolated people, physically and socially, from the places where they conduct their daily business.

Today, zoning and subdivision regulations essentially mandate sprawling suburban development. Research has shown that there is a link between the land use patterns determined by current zoning and subdivision ordinances and declining environmental quality. Data also suggest that traditional urban typologies, such as small towns and villages, may be environmentally benign relative to those created by current policies. By reducing the amount of land per person, or urban land use ratio (ULUR), from current suburban values of 3 to 20 acres per person (which does not represent solely the land that a person lives on, but rather the amount of land in urban use, including roads and parking lots), to lower, more traditional values of 0.1 to 0.05 acres per person, population densities will increase (which research has shown has no effect on the quality of life) and thereby accommodate the projected population increase in a smaller proportion of the coastal landscape.

Research on New Urbanist and neotraditional landscape architecture points to the profitability of these designs and suggests that they are capable of outperforming conventional suburban typologies in the market. This is not to say that all new developments should look like nineteenth-century Victorian communities. However, by restructuring land use policies to allow reduced ULURs, reduced road widths to traditional dimensions, and protection of natural riparian buffers (often done out of necessity in the past), as well as by encouraging new development in areas with infrastructure already in place, positive environmental and fiscal outcomes should be realized.

A variety of conservation tools, including conservation easements, the purchase and transfer of development rights, and the creation of urban service or growth boundaries have been used successfully to protect existing nonurban parcels and landscapes. To be successful, however, conservation efforts must involve all stakeholders, particularly property owners and developers. Incentives that reward development that the public considers appropriate in preapproved locations reduce risks and costs to developers and create positive outcomes for communities.

Although benign alternative landscape architectures exist at the scale of the subdivision development, application of sustainable design principles to larger landscapes, such as the county and region, have generally been lacking. At these scales one must consider a combination of alternative approaches to achieve preservation of landscapes, rather than parcels. Standard, publicly supported conservation techniques such as the purchase of development rights are usually inadequate to protect more than a small portion of a landscape under development pressure. A strategy (referred to as the Hamlet Approach) that involves government, developers, and landowners may provide a mechanism for protecting rural landscapes while encouraging economic growth and development and maintaining the value of the land for owners of large parcels. First, the local jurisdiction must identify some number of 130- to 500-acre development districts that will be developed at traditional densities (ULURs of 0.1–0.07). The number of districts depends on expected growth. Developers purchasing parcels in the districts must then purchase development rights from local landowners outside of the districts and transfer them into the districts. The transferred development "credits" can be used to build at the higher allowable density. This transfer of development rights approach preserves land in proportion to the development pressure. Coincidental to the process, infrastructure extension is constrained, which reduces costs to both developers and taxpayers.

Current demographic trends indicate that the conversion of land to urban use will not abate any time soon. Current urban development patterns are causing environmental and ecological deterioration and loss of cultural identity at unprecedented rates in many parts of the United States. Alternatives to current development policies exist and must be considered if there is to be any chance at preserving the functioning of ecosystems and the integrity of our nation's regional cultures. Opportunities to become educated in the methods of sustainable development are available to decision-makers and to the public at large. The failure of communities to incorporate these design principles into local decision-making will invariably reduce the quality of life while increasing the costs of community services to citizens.

2.1 Introduction

For approximately a century after the Civil War, the American Southeast experienced one of the lowest rates of regional economic development in the nation (DeVoe and Kleppel 1995). Emigration, driven by economic stagnation, and the concentration of property in the hands of a few individuals and corporations combined to produce one of the most sparsely populated coastlines in America. In the mid-1970s, however, the coastal real estate market exploded. Today, the Southeast is among the fastest-growing regions in the country (Culliton et al. 1990). The U.S. Census Bureau projects that approximately 11 million people, some 14 percent of the post-World War II baby boom generation, will immigrate to the Carolinas and Georgia between 1995 and 2025 (Kleppel and DeVoe 2000). Fifty to 75 percent of the in-migrating population will reside on the coastal plain (DeVoe and Kleppel 1995). As a result, ecosystems will be degraded and traditional cultures and economies will be displaced (Van Dolah et al. 2000).

In this chapter we consider some of the causes and consequences of the remarkable change in land use patterns that has occurred in the coastal Southeast. We argue that while development represents the proximate cause of cultural and environmental change, policies that govern land use and development are the ultimate forcing functions of the changes that take place.

We will take a look at the history of coastal development in the Southeast, particularly in South Carolina, and consider the likely trajectories of future development. We will examine some of the federal and local policies that drive these trajectories and the typologies they create. Finally, we will suggest specific modifications to existing policies, particularly those governing zoning and subdivision design, that we believe will lead to culturally and environmentally benign changes in urban typology, while impacting neither the rate of population growth nor the expected benefits of economic development.

2.2 Trajectories of Coastal Population Growth and Development

2.2.1 A Brief History of Development in the Coastal Southeast

Following the Civil War and the failure of the plantation system, large parcels of land were acquired by silvaculture and paper interests, and by wealthy individuals seeking the idyllic seclusion of the region's coasts and barrier islands. Elegant vacation homes, hunting lodges, private estates, and posh resorts replaced cotton and rice plantations. Bernard Baruch acquired the 15,000-acre Hobcaw Barony outside of Georgetown, South Carolina, for his vacation home. R. J. Reynolds established a private hunting preserve on Sapelo Island, off the coast of Brunswick, Georgia. And Jekyll and St. Simons Islands off the Georgia coast became the sites of splendid Victorian and Arts and Crafts era "cottages" and hotels. Interestingly, "privatization" of the coast obviated the possibility of intensive development and, quite unintentionally, preserved many of the region's unique coastal ecosystems. Today, most of these private preserves are state or national parks.

Creation of the Cape Romaine National Seashore in 1932, and the Francis Marion National Forest in 1936, along with the protection of large coastal tracts, including 135,000 acres in the watersheds of the

Ashepoo, Combahee, and Edisto Rivers—the ACE Basin—in 1991, has reduced the amount of developable land along the shoreline of South Carolina considerably (DeVoe and Kleppel 1995). The population of the South Carolina coastal plain will increase by more than 500,000 between 1995 and 2025, and will only be accommodated by tracking inland along rivers, lakes, and major highways (Allen and Lu 2003).

2.3 Land Use Policies and their Effects on Coastal Development

2.3.1 Risk, Real Estate, and the National Flood Insurance Program

Historically, the trajectories of residential development in the coastal zone reflect regional adaptations to risk. The assumption of risk has always been a mitigating element on land use decisions, and early low-lying structures built in close proximity to tidal rivers and coastal dunes illustrate the adaptation to risk. Small fishing cabins and cottages were built recognizing the possibility that their very location put them in danger of destruction (by hurricanes or floods) and that they might need to be replaced. Great homes and hunting clubs, such as Carnegie's Dungeness mansion on Cumberland Island and the Jekyll Island Club, both off the Georgia coast, though grand in scale, reflected the same sense of risk management. The properties were self-insured by great wealth, and siting considerations included risk reduction. Thus, the Jekyll Island Club sits almost a mile inshore of the dune line.

When risks can be externalized, and when associated costs can be reduced or borne by others, perversions in markets, as well as in individual behaviors, occur. This is, perhaps, nowhere better illustrated than by the influence of the National Flood Insurance Program (NFIP) of 1968, the accompanying Flood Disaster Protection Act of 1973, and the Flood Insurance Reform Act of 1994, which required borrowers of federally backed mortgages to purchase subsidized flood insurance coverage (Myers 1995).

Until the NFIP became available, banks usually refused to grant mortgages for beachfront home construction without additional backing with low-risk securities. It was only after the federal government assumed the risk private insurers were unwilling to undertake, that the coastal real estate market boomed.

Kriesel and Landry (2004) examined participation in the NFIP for coastal zone properties. They found that, unlike the programs in river-basin floodplains, where participation has been historically low, coastal participation was significant, with almost 50 percent of homes estimated to be part of the program. Of those responding to the survey, more than 82 percent reported having coverage by the NFIP. The authors weighted that response downward to fit "known" insured property distributions. Within their sample the leading predictor of participation was the requirement of a lender for such subsidized coverage. Again, the lenders' requirement that the property owner purchase flood insurance reflects a reduction in risk and therefore opens to development housing markets that were previously considered too risky for completely internalized cost assumption.

While it was the intent of Kriesel and Landry to examine price elasticity and individual decision criteria for participating in the NFIP, the evidence of risk minimization required by lending institutions is instructive and further suggests that the NFIP may be an extremely influential driver of rapid coastal residential development.

Figure 2.1 illustrates the historic land parcel transaction activity in the Murrell's Inlet area of Georgetown County, South Carolina. Land parcel transactions are defined as the sale or exchange of existing parcels, and also as the subdivision, sale, and transfer of larger parcels. Between 1908 and the early 1970s there was little variation in activity, with only modest increases in parcel transactions, principally as the result of infrastructure (e.g., road and bridge) development. After the 1970s, parcel transaction exhibits exponential growth. An explanation for this growth is the entry of Georgetown County into the NFIP and the subsequent reduction of risk in this market. The idea of risk reduction is strengthened by the dip in parcel transactions in the early 1990s. The dip has been attributed to the implementation of the South Carolina Beach Management Act (BMA) and the state's post-Hurricane Hugo (September 1989) refusal to allow construction on highly eroded beach lots, typified by those owned by David Lucas on a Charleston County barrier island (Butler 1993). The creation of setbacks under the BMA, advanced by South Carolina as an appropriate police zoning action, was contested by Lucas as a taking under provisions of the Fifth Amendment to the U.S. Constitution. That case, settled in the U.S. Supreme Court in 1992 with Lucas the victor, was a landmark in advancing the so-called property rights agenda. Access to subsidized federal insurance is encompassed by the takings concept.

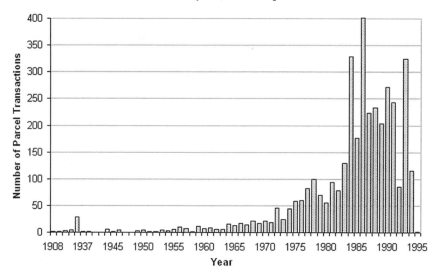

Fig. 2.1. Land parcel transactions in the Murrells Inlet area of South Carolina, 1908–1995

Beaufort County, South Carolina, with such notable sites as Hilton Head Island, is among the fastest-growing retirement and resort locations in the nation (Figure 2.2). Hilton Head was a principle driver for parcel creation in the period from 1960 to 1980, with peaks occurring in conjunction with infrastructure improvements. As in the Murrells Inlet area, parcel transaction in Beaufort County exhibits a depression in the early 1990s, i.e., the post-Hurricane Hugo period, and then resumes strong growth with the subsequent clarification of coastal zone management and property rights issues.

Although the NFIP is not the sole driver of growth in coastal South Carolina, it would appear to make a clear contribution to the pattern of development. Even Gilbert White (1994), who is considered the father of flood plain management, contends that the program was not intended to have the consequences that it has. The NFIP was supposed to be a source of retreat from the coastal and floodplain zones—not a property right to allow building where even common sense would suggest we not go.

Helvarg (2003) is less kind regarding the influence of the NFIP. He comments:

> Barrier islands are like geology on amphetamines. Unarmored, they tend to move, by the decade, year, season, sometimes in a

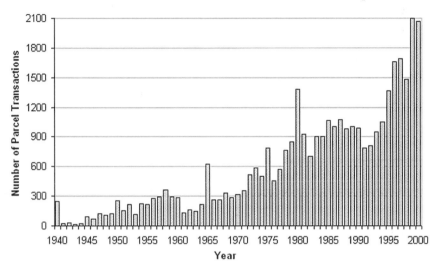

Fig. 2.2. Land parcel transactions in Beaufort County, South Carolina, 1940–2000

single stormy day. It's a natural process that can be strikingly beautiful, provided you haven't just closed on a multimillion-dollar beachfront dream home.

Unfortunately, more and more wealthy people in the United States are doing just that—moving to or buying second homes in places like East Hampton, Hilton Head, Sea Island, Ocean Reef or Captiva. Other folks are buying high rise condos in Ocean City, Myrtle Beach, Gulf Port, Coronado and Honolulu, or else flocking like lemmings to new housing developments built on filled-in salt marshes and flood plains all along the coast.

And while lemmings don't actually jump off cliffs and drown themselves in the sea, if they did there would undoubtedly be a number of government programs offering them flight insurance and full-coverage for any water damage....

There are, however, more sensible models for living safely by the shore in the heavy weather to come. When Hurricane Opal hit the Florida panhandle in 1995, towns like Destin and Dune-Allen were devastated, but the town of Seaside—a recently constructed village of 280 old-Florida-style frame homes set back behind the dunes—came out unscathed.

Aside from siting its homes behind the beach's protective sand dunes, Seaside also used tough building standards designed to resist 150-mile-per-hour winds, sank its foundation pilings deep into the ground, and planted lots of native trees and grasses to secure the dunes and buffer the houses. Also rejected were cheap building materials like vinyl and strandboard siding. "If it didn't exist before 1940, we didn't feel it was proven," says Robert Davis, the founding developer of Seaside.

"It used to be everyone built well back from the beach," he adds. "That was before federal flood insurance made stupidity feasible."

2.3.2 Zoning, Land Use, and the Geometry of Communities

Federal policies such as the NFIP drive behaviors and outcomes at the local level, where the overwhelming majority of land use decisions are made (Dale et al. 2000). However, as White (1994) pointed out (above), the behaviors and outcomes that occur as a result of these policies are not always those intended. The exemplar of unintended consequences in land use policy is single-purpose zoning and its corollary ordinances for parcel subdivision and infrastructure development.

The concept of zoning emerged in Europe during the nineteenth century. Urban planners realized that the siting of factories in residential neighborhoods threatened the public well-being and they sought to separate industrial from residential areas within cities. In the United States, the State Standard Zoning Enabling Act (SZEA), of 1926, extended that concept to single-purpose districting, which endeavored to cluster consistent land uses and activities into distinct zones and to exclude inconsistent or nonconforming uses. An unintended consequence of this policy during the past 60 years has been the increasing separation of people from the businesses, services, and institutions they regularly use (Salkin 2002).

In the 1940s and 1950s, the automobile emerged (through pressure applied in Washington by oil and automobile companies) as the centerpiece of a new perception of mobility in America (Katz 1994). Construction of roads and highways became nearly the singular focus of federal transportation policy. Public transportation was ignored and its infrastructure fell into disrepair. The quality of service became undependable and, predictably, ridership disappeared (Caro 1974). Since the 1940s, the coupling of single-purpose zoning with policies promoting highway development and personal vehicle use (over public transportation) has rescaled the American ambit to the automobile, as is most obviously expressed by

Table 2.1. Urban Land Use Ratios (Acres in Urban Use/Person) Characteristic of: A. Traditional (Pre-1945) and B. Modern and Suburban (Post-1945) Development Styles.

Location	Ratio	Reference or data source
A. Traditional		
Brunswick, ME	0.04	City of Brunswick, ME
Boston, MA	0.05	City of Boston, MA
Baltimore, MD	0.10	City oof Baltimore, MD
Troy, NY	0.14	Fabozzi, 2002
Charleston, SC	0.05	City of Charleston, SC
B. Modern suburban		
Beaufort County, SC	3.0	Allen and Lu, unpublished
Guilderland Township, NY	5.0	Fabozzi 2002
Town of Clifton Park, NY	7.0	Fabozzi 2002
Berkeley County, SC	7.0	BCD-COG 1996
Dorchester County, SC	8.0	BCD-COG 1996
Albuquerque, NM	6.9	City of Albuquerque, NM
Central New York State	10.0	Pendall 2003
Los Angeles, CA	20.7	City of Los Angeles, CA

the changing relationship between people and the urban landscape. This relationship is expressed by the urban land use ratio, or ULUR, which is simply the amount of urban land/person in a jurisdiction. A drastic change in the ULUR in American communities occurred after World War II, coincident with the passage of the G.I. Bill, the development of the modern suburban typology and the emergence of the automobile as the mainstay of American mobility. In urban areas that experienced their major development prior to about 1945, ULURs are generally below 0.1 (Table 2.1A). Urban areas that developed after 1945 have ULURs ranging from 3 to 20 (Table 2.1B). The per capita rate of urbanization today is about two orders of magnitude higher than it was prior to 1945.

2.3.3 Unintended Consequences of the Standard Zoning Enabling Act

Single-purpose zoning has evolved beyond its initial intent of removing heavy industry from residential neighborhoods. Today, it segregates people by social and economic class and has produced a mind-set that considers it inappropriate to live within walking distance of the businesses and services one uses on a regular basis. Since the late 1980s architects and planners, in growing numbers, have come to consider the infrastructure, design, and spatial-scaling elements created by our modern zoning and subdivision ordinances to be aesthetically unappealing, spatially wasteful, poorly proportioned, and sometimes, even sociopathic (Kunstler 1993; Fodor 1999; Duany et al. 2000).

A considerable body of research, accumulated during the past 30 years, indicates that receiving-waters draining communities developed according to modern zoning and subdivision standards exhibit excessive nutrient enrichment, eutrophication, toxic contamination and the loss of ecosystem functionality (Valiela and Teal 1979; Limburgh and Schmidt 1990; Scheuler 1994; Nixon 1995; Fulton et al. 1996; Hoss and Engel 1996; Lerberg et al. 2000; Van Dolah et al. 2000; Bowen and Valiela 2001; Valiela and Bowen 2002; Holland et al. 2004; Kleppel et al. 2004). Nonetheless, few studies have attempted to associate these impacts with specific urban typologies. Recent work, suggests that traditional urban typologies (e.g., small towns, villages) may be environmentally benign relative to modern suburbs (Kleppel et al. 2004). Kleppel et al. (this volume) predict that damage to tidal creek-benthic ecosystems, likely to result from rapid population growth and development in South Carolina's coastal watersheds, may be substantially reduced by lowering urban land use ratios from those typical of modern suburbs, to values characteristic of traditional towns and villages. Although runoff to tidal creeks increases in proportion with the amount of urbanized land in coastal watersheds (Holland et al. 1996; Holland et al. 2004; Lerberg et al. 2000), the type of development in a watershed may influence susceptibility to nonpoint source contamination (Kleppel et al. 2004; Shirer 2004). Wetlands that drain suburbanized watersheds receive larger amounts of nonpoint source runoff than wetlands in watersheds where the principal form of development is a small town, village, or hamlet, even when approximately the same number of people lives in each watershed (Kleppel et al. 2004). At least three factors—the ULUR, the average width of vegetative buffers, and the amount of impervious surface within the watershed—differentiate traditional from suburban land use patterns and appear to influence the impacts of urban development on ecosystem functioning (Kleppel et al. 2004).

Modern urban designs based on traditional typologies, with mixed commercial and residential districts and greater capacity for walking, are being created throughout the United States by rezoning and over-lay zoning (Duany et al. 2000). This neotraditional typology is competitive in the market (Eppli and Tu 1999). In the past five years, there has been a noteworthy increase in neotraditional, mixed-use subdivision development in the Carolinas, often with concomitant reductions in imperviousness and even attention to traditional architecture. Similarly, revitalized urban neighborhoods and small towns are attracting the attention of home buyers.

2.3.4 Reducing Development Pressure on Large Parcels of Land

One of the frequent outcomes of single-purpose zoning is "hopscotching" suburban development. Unlike the suburbs of the nineteenth and early twentieth centuries, which grew as neighborhoods along the edges of existing urban centers, hopscotching occurs as isolated pockets of growth greatly displaced from the core. Promoted by planners and engineers after World War II, who envisioned America as a society of commuters (Katz 1994), and by school districts and commercial interests seeking inexpensive land (understanding that the infrastructure will be publicly subsidized), hopscotching has become a principal cause of urban decay, loss of family farms, and landscape fragmentation, which severely impacts biological diversity. Today, more than two million acres of land in the United States are urbanized each year (Fodor 1999). More than half of that is on the coastal plain.

Virtually all of the developable land in the coastal Southeast is privately owned, and efforts to manage the rate and impacts of urbanization must target the owners of undeveloped large parcels. Among the most widely used conservation instruments are easements, the purchase or transfer of development rights (PDR, TDR), and the creation of urban growth and service boundaries (UGB, USB) (Kleppel 2002). In Lancaster County, Pennsylvania, agricultural zoning and easements have protected some 14,500 acres of fertile, productive farmland (Daniels and Bowens 1997; Daniels 1999). Conservation easements in the ACE Basin of South Carolina have similarly protected large amounts of land in this sensitive estuarine system.

Under such an approach, landowners restrict development of their land and may receive fee simple compensation or some other benefit, such as a tax reduction. The easement may be held and administered by a land conservancy, some government authority, or a combination of private and public entities. A landowner also may sell the development rights attached to his property (a PDR) to a public or private interest, or to a consortium, in return for cash and/or tax benefits. The landowner forfeits future development options but retains title to the property, managing it for traditional rural uses. Use of the development rights by the purchaser(s) is legally constrained. Because PDRs involve willing sellers and willing buyers, property rights remain protected.

Various mechanisms are used to fund PDR programs, including sales taxes, real estate fees, and property taxes. PDR programs are capable of constraining hopscotch and sprawling development and thereby tend to reduce tax burdens associated with the extension of infrastructure and urban services. Beaufort County, South Carolina, has already established a PDR program, and Charleston County may soon follow suit. Unfortunately, the number of property owners interested in PDR programs almost always exceeds the availability of funds.

Transfer of development rights (TDR) programs, dependent on private sector investment, may be better able to protect land in proportion to development pressure. As such, TDRs may be more effective conservation instruments than PDRs because the investment in conservation can be made by the developer, who is rewarded by gaining greater access to the market within a sanctioned development district (DD). Urban growth and service boundaries beyond which publicly supported infrastructure is not extended, and the designation of development districts near existing infrastructure (obviating the need for new, publicly financed infrastructure development) have contributed to urban and neighborhood revitalization and the protection of open space and farmland in Oregon, Kentucky, the Chesapeake Bay watershed, Maine, and Maryland. Not surprisingly, UGBs, USBs, and DDs have also been effective at controlling the rate of tax growth.

Alternatively, incentives in the permitting process that encourage development that the community considers acceptable can replace antagonism between developers and residents with a sense that development will have a positive outcome (Duany et al. 2000). At public workshops, stakeholders reach agreement on the attributes and locations of "acceptable" development districts. Developers are provided with this information (usually as a simple checklist of development standards and a list of "shovel-ready" sites in prezoned development districts). Permitting is expedited if the developer complies with the program. The time saved can significantly improve the developer's bottom line. Alternatively, the applicant may choose to develop in the conventional way, in which case, a rezoning petition is submitted and the rezoning process begins. Rezoning may take months to years, and may end in failure.

2.4 Fixing the System—The Hamlet Approach

In this chapter we have suggested that the NFIP and property rights advocacy are in large part responsible for the explosive rise in development along the Southeast coast since the mid-1970s. We have predicted that during the next 30 years, the trajectory of development on the coastal plain will track inland, along rivers, lakes, and major highways. We have shown that the present distribution of development and the form it takes on the landscape are defined by local ordinances governing zoning and the subdivision of property as well as policies supporting development of various kinds of infrastructure (e.g., highways) over others (e.g., rails). We have suggested that the urban typologies created by these ordinances often have negative environmental and social impacts. Finally, we have argued that marketable alternative development styles, legal instruments, and incentive techniques exist, and that together, they can be used to create relatively benign urban typologies.

We conclude this chapter by describing how some of these alternative typologies and conservation tools can be applied, emphasizing that no approach will be effective alone and that no specific combination of approaches will be effective in every case. Furthermore, one can expect that developers and realtors, local government officials, and citizens who believe that the alternatives will limit their rights or reduce their quality of life, will likely resist changes to existing land use policies. In any effort to create and implement a sustainable development strategy, the developers, the citizenry, or both must be firmly on board (or at least disinterested). To gain the support of these stakeholder groups, it is crucial to demonstrate that their interests and quality of life will be protected. To this end, education of the public with regard to the issues and the honest sharing of information is crucial in generating support and alleviating concerns about development within a community.

The landscape architecture of the neotraditional subdivision development has described extensively in the literature (Arendt 1992, 1999; Duany et al. 2000). However, issues of scale, spatial distribution, and density within the larger jurisdiction (the town or county), remain largely unresolved. For example, decisions about the number and density of subdivisions within a jurisdiction and what to do with parcels that will not be developed are not addressed at the typology scale and often are not considered or are inadequately addressed in ordinances and municipal codes. A failure to address these issues may simply alter the architecture of sprawl and will most assuredly bring out the property rights constituency. Such efforts are usually doomed before they start.

The best way to achieve preservation of large parcels is to create a strategy that will allow buy-in by both landowners and developers. Creating a system of easements and PDRs may satisfy landowners, but it will leave developers, realtors, and often, other influential players (e.g., moneylenders, remote public works authorities seeking to extend their systems) out of the matrix, and may result in costly and contentious disputes. Further, in most cases public and conservancy funds are insufficient to protect more than a fraction of the developable land.

We propose here an alternative based on private sector investment. We call this concept the Hamlet Approach. A hamlet differs from a village or a town in that villages and towns are government jurisdictions; a hamlet is not. Thus, while the power and tax structures of a community would be altered by creating villages or towns, they will not change by creating hamlets. A hamlet is simply a traditional multi-use urban district (see Kleppel et al. 2004) within a particular jurisdiction. A hamlet may be a development district separate from an existing urban area, or it can be attached to an existing urban system, such as a city, in which case it becomes a neighborhood (see Duany et al. 2000). Hamlets themselves, however, will not create sustainable landscapes. A critical part of the Hamlet Approach is to appeal to both developers and landowners by using conservation instruments, particularly TDRs, to create or enhance value in both developable (the hamlets) and nondevelopable districts (the large parcels and landscapes surrounding hamlets).

TDR programs require sending and receiving zones, from which and to which development rights move. Sending zones are usually large, privately owned parcels, such as farms and forests. Receiving zones, the hamlets, are some number of mixed-use, urban developments created by the jurisdictional authority (e.g., the county) and developed by several developers with expertise in residential, commercial, and public sector architecture and real estate. The number of hamlets is dependent on several factors, and we demonstrate how the number of hamlets influences urban land use and large parcel protection below. Hamlets would be located to avoid impacts to cultural and natural resources. They would be separated from one another by legally specified distances, as currently exist for towns under South Carolina law. Each hamlet would receive water and sewerage from the jurisdiction, but taps to water and sewer lines would be prohibited outside the hamlet. The jurisdiction would invest in the hamlet by providing additional services (e.g., police and fire substations, library annexes).

Planners and landscape architects suggest that in a hamlet, the average person should be able to walk from one side of the community to the other in about ten minutes, or to go from the edge of the community to its center in five minutes. This would constrain hamlets to a diameter of about a half-mile, which, if the hamlet is envisioned as a circle, would constitute an area of 0.20 square miles or 125.9 acres. Alternatively, it seems unlikely that the walkability of a community would be compromised by a doubling of its diameter to a mile (a 15- to 20-minute walk from end to end). This would have a large impact on area, increasing the size of the development district by a factor of four, to 0.79 square miles or 502.7 acres. The result of enlarging the area of the hamlet would be to reduce the number of hamlets needed to accommodate some amount of expected population growth and would probably help to defray some infrastructure costs. Smaller hamlets might be desirable to real estate interests, as well as to organizations (banks) that will sell bonds for infrastructure. Larger hamlets would appeal to public authorities charged with infrastructure development and to developers who bear some of the costs of building the infrastructure.

Population densities might vary from one hamlet to the next, but hamlets would seek traditional average ULURs lower than 0.1 (Table 2.1A). ULURs well below 0.1 do not detract from marketability or perceived quality of life (Arendt 1992). The ULUR of Charleston, South Carolina, for instance, is 0.05. Of course, not everyone would live at the same density; traditional communities permit a variety of residential densities, with highest densities closest to (or within) desirable shopping or public districts or near other urban services (e.g., the village green or a public garden). Lower densities might occur near the outer edge of the hamlet, or near certain other public landscapes, such as a golf course. The jurisdiction would logically provide more services to hamlets with higher average densities.

Using land use and land cover data (Table 2.2) and U.S. Census Bureau growth projections for Beaufort County, South Carolina (the fastest-growing coastal county in the state between 1990 and 2000), we generate a series of scaling scenarios (Table 2.3) for hamlets with different length standards (diameters of 0.5 and 1.0 mile) and ULURs

Table 2.2. Land Cover Data in Beaufort Country, South Carolina. A. Land Covers and Uses. B. Acreages of Developable and Undevelopable Areas of the Country.

A. Land Cover Classes and Acreages				
Land Cover Class	Acres	Land Cover Class	Acres
High density urban	1008	Freshwater marsh	1169
Low density urban	790	Deciduous wet forest	4534
Evergreen forest	85610	Evergreen wet forest	31010
Deciduous forest	4378	Bottomland hardwood	7314
Mixed forest	1991	Wet scrub/shrub	44164
Open	38355	Dry scrub/shrub	9929
Water	218145	Sand Beach	1524
High marsh	55349	Barren	2890
Low marsh	78767	Total	586926

B. Developable and Undevelopable Areas in Beaufort County		
Land Cover Class	Acres	Percent of Total
Water/wetlands	433138	73.8
Existing urban	1798	0.3
Total Developable	151990	25.9

(0.10 and 0.07). The decision to use Beaufort County in this example was arbitrary and implies no special interest on our part.

At the time that the land cover and land use data were compiled (2000), nearly 76 percent of the land in Beaufort County was classified as wetland. Less than 1 percent of the county was already developed. Slightly more than 24 percent of the land in Beaufort County, about 152,000 acres, was nominally available for development. Between 2000 and 2025, the population of Beaufort County is expected to increase by 62,239 people.

As is evident from Table 2.3, the quadrupling of hamlet area that occurs with a doubling of its diameter results in a quadrupling of the hamlet population and a 75 percent difference in the number of hamlets in the county, for a given population density (ULUR). Furthermore, regardless of the number of hamlets created, the amount of land developed is reasonably constant, the difference (<1000 acres) being due to differences in the ULUR, not the diameter. The amount of land conserved by this approach is about the same in all cases and exceeds 95 percent. Although we have not factored the spatial requirements of

Table 2.3. Project Outcomes of Three Implementation Scenarios of the Hamlet Approach in Beaufort Country, South Carolina. Predictions are Based on Different ULURs and a Population Increase of 62,239 People Between 2000 and 2025.

Hamlet Attribute	Scenario			
	1	2	3	4
Diameter of hamlet (mile)	1.0	1.0	0.5	0.5
Average ULUR[a]	0.10	0.07	0.10	0.07
Hamlet area (acres)	503	503	126	126
No. of hamlets in county	12	9	50	35
Maximum population	5027	7181	1259	1799
Maximum no. of units[b]	1676	2393	420	600
Acres developed in county	6032	4524	6224	4407
Acres conserved in county	145958	147465	145766	147583
Percent developable land conserved	96	97	96	97

[a] ULUR= Urban land use ratio (acres/person.)
[b] Maximum number of units = maximum number of people/3.

infrastructure into the calculations, the scenarios nonetheless provide a reasonable sense of the development scales and conservation implications of the Hamlet Approach.

It bears mentioning that the Hamlet Approach does not attempt to reduce growth or development within the jurisdiction. It does not affect the number of people who will be looking for places to live. It does, however, reduce infrastructure and overhead costs both for developers and the public, while simultaneously conserving large amounts of land.

Most importantly, development of hamlets is underwritten by private sector investment, through the purchase of transferable development rights from land outside the development district (hamlet), by the developer. The proposed TDR program provides the capital to ensure that the land outside of the development districts retains its value. The development rights purchased from a property owner outside a hamlet (see Daniels and Bowers 1997) are used to increase unit density (to ULUR specifications) on the developer's parcel within the hamlet.

The density permitted per acre of transferred development rights is determined by jurisdictional formula (Kleppel and Mapes unpublished ms.). The limited funds available through PDR programs can now be directed at protection of parcels with public value, such as historic sites or ecologically sensitive landscapes.

2.5 Conclusion

Human demographics ensure that urbanization in the coastal Southeast will not abate any time soon. Conservative estimates of growth trajectories and projections of the impacts of development portend severe environmental and cultural consequences if current land use patterns persist. It is clear however, that population growth alone drives neither the scale nor the impact of urbanization. Rather, local and regional policies that regulate the location and typology of development ultimately determine impacts on both natural and cultural resources. Environmentally benign typologies that are competitive in the market already exist. It becomes incumbent on local decision-makers and their advisers to become familiar with the design principles and conservation techniques that we have described here. The failure of local governments to incorporate these principles into policy leads to documented failures in economy (Burchell et al. 2000; AFT 2001), public health (Lopez 2004), and environmental quality (Kleppel et al. 2004). Decision-makers must embrace the principles of benign urban design or they must be prepared to accept responsibility for the consequences.

Acknowledgments

We are grateful for technical assistance provided by Kang Shou Lu and Jeffrey Mapes. Partial support for this project was provided as part of NOAA's Land Use-Coastal Ecosystem Study through grant number NA960PO113 from the South Carolina Sea Grant Consortium.

References

Allen, J. and K. Lu. 2003. Modeling and prediction of future urban growth in the Charleston region of South Carolina: A GIS-based integrated approach. *Conservation Ecology* 8:2. 20p. http://www.consecol.org/vol8/iss2/art2/.

American Farmland Trust (AFT). 2001. The cost of community services. Technical Paper. AFT, Washington, DC. 1p.

Arendt, R. 1992. Rural by Design. American Planning Association, Chicago. 441p.

Arendt, R. 1999. Growing Greener. Island Press, Washington, DC. 236p.

Bowen, J. L. and I. Valiela. 2001. The ecological effects of urbanization of coastal watersheds. Historical increases in nitrogen loads and eutrophication of Waquoit Bay estuaries. *Canadian Journal of Fisheries and Aquatic Science* 58:1489–1500.

Burchell, R. W., A. Downs, S. Mukherji, and B. McCann. 2000. The costs of sprawl. National Academy of Sciences Press, Washington, DC. 100p.

Butler, H. N. 1993. Regulatory takings after Lucas. *Regulation: The Cato Review of Business & Government* 3. The Cato Institute, Washington, DC. p. http://www.cato.org/pubs/regulation/reg16n3g.html.

Caro, R. A. 1974. The Power Broker. Vintage, New York, NY. 1246p.

Culliton, T. J., M. A. Warren, T. R. Goodspeed, D. G. Remer, C. M. Blackwell, and J. J. I. McDonough. 1990. Fifty years of population change along the nation's coasts 1960–2010.. U.S. Department of Commerce, National Oceanic and Atmospheric Administration (NOAA), Rockville, MD. 41p.

Dale, V. H., S. Brown, R. A. Haeuber, N. T. Hobbs, N. Huntley, R. J. Naiman, W. E. Riebsame, M. G. Turner, and T. J. Valone. 2000. Ecological principles and guidelines for managing the use of land. *Ecological Applications* 10:639–670.

Daniels, T. 1999. When City and Country Collide. Island Press, Washington, DC. 361p.

Daniels, T. and D. Bowers. 1997. Holding Our Ground: Protecting America's Farms and Farmland. Island Press, Washington, DC. 334p.

DeVoe, M. R. and G. S. Kleppel. 1995. The South Atlantic Bight Land Use-Coastal Ecosystem Study. South Carolina Sea Grant Consortium, Charleston, SC.

Duany, A., E. Plater-Zyberk, and J. Speck. 2000. Suburban Nation. North Point Press, New York, NY. 293p.

Eppli, M. J. and C.C. Tu. 1999. Valuing the new urbanism. Urban Land Institute, Washington, DC. 86p.

Fabozzi, T. M. 2002. Quality Region Task Force Report. Capital District Regional Planning Commission, Albany, NY. 25p.

Fodor, E. 1999. Better, Not Bigger. Island Press, Washington, DC. 182p.

Fulton, M. H., G. T. Chandler, and G.I. Scott. 1996. Urbanization effects on a southeastern U.S.A. bar-built estuary, pp. 477–504. In, Vernberg, F. J., W. B. Vernber, and T. Siewicki (eds.), Sustainable Development in the Southeastern Coastal Zone. University of South Carolina Press, Columbia.

Helvarg, D. 2003. Coasts at risk: Coastal sprawl and the shore. *The Multinational Monitor* 24. 1p. http://www.multinationalmonitor.org/mm2003/03september/sept03corp3.html.

Holland, A. F., et al. 1996. The Tidal Creek Project. Interim Report. Charleston Harbor Project, South Carolina Department of Health and Environmental Control, Charleston. 229p.

Holland, A. F., D. M. Sanger, C. P. Gawle, S. B. Lerberg, M. S. Santiago, G. H. M. Riekerk, L. E. Zimmerman, and G. I. Scott. 2004. Linkages between tidal creek ecosystems and the landscape and demographic aspects of their watersheds. *Journal of Experimental Marine Biology and Ecology* 298: 151–178.

Hoover, H. 1926. Standard State Zoning Enabling Act. U.S. Department of Commerce, Washington, DC. 9p.

Hoss, D. E. and D. W. Engel. 1996. Sustainable development in the southeastern coastal zone: Environmental impacts on fisheries, pp. 121–140. In, Vernberg, F. J., W. B. Vernberg, and T. Siewicki (eds.), Sustainable Development in the Southeastern Coastal Zone. University of South Carolina Press, Columbia.

Katz, P. 1994. The New Urbanism. Toward an Architecture of Community. McGraw-Hill, New York, NY. 245p.

Kleppel, G. S. 2002. Urbanization and environmental quality: Implications of alternative typologies. *Albany Law Environmental Outlook Journal* 8:37–64.

Kleppel, G. S. and M. R. DeVoe. 2000. The people factor, pp. 32–43. In, Davis, J. (ed.), A Sense of Place. South Carolina Wildlife, South Carolina Department of Natural Resources, Columbia.

Kleppel, G. S., S. Madewell, and S. E. Hazzard. 2004. Responses of emergent marsh wetlands in upstate New York to variations in urban typology. *Ecology and Society* 9(5):1. 18p. http://www.ecologyandsociety.org/vol9/iss5/art1.

Kriesel, W. and C. Landry. 2004. Participation in the National Flood Insurance Program: An empirical analysis for coastal properties. *Journal of Risk and Insurance* 71:405–420.

Kunstler, J. H. 1993. The Geography of Nowhere. Simon and Schuster, New York, NY. 303p.

Lerberg, S. B., A. F. Holland, and D. M. Sanger. 2000. Responses of tidal creek macrobenthic communities to the effects watershed development. *Estuaries* 23:838–853.

Limburgh, K. E. and R. E. Schmidt. 1990. Patterns of fish spawning in the Hudson River watershed: Biological response to an urban gradient? *Ecology* 71:1238–1245.

Lopez, R. 2004. Urban sprawl and the risk of being overweight or obese. *American Journal of Public Health* 94:1574–1579.

Myers, M. F. 1995. The National Flood Insurance Program as a nonstructural mitigation method. Natural Hazards Research and Application Information Center, University of Colorado, Boulder. 13p.

Nixon, S. W. 1995. Coastal eutrophication: A definition, social causes and future concerns. *Ophelia* 41:237–249.

Pendall, R. 2003. Sprawl without growth: The upstate paradox. Brookings Institution, Washington, DC. 11p.

Salkin, P. E. 2002. Smart growth and sustainable development: Threads of a national land use policy. *Valparaiso University Law Review* 36:381–412.

Scheuler, T. R. 1994. The importance of imperviousness. *Watershed Protection Techniques* 1:100–111.

Shirer, R. 2004. Effects of landscape disturbance on freshwater emergent wetlands in the Hudson Valley, New York. Master's thesis, State University of New York at Albany. 72p.

Valiela, I. and J. L. Bowen. 2002. Nitrogen sources to watersheds and estuaries: Role of land cover mosaics and losses within watersheds. *Environmental Pollution* 118:239–48.

Valiela, I. and J. M. Teal, 1979. The nitrogen budget of a salt marsh ecosystem. *Nature* 280:652–656.

Van Dolah, R. F., D. E. Chestnut, and G. I. Scott. 2000. Baseline assessment of environmental and biological conditions in Broad Creek and the Okatee River, Beaufort County, South Carolina. National Ocean Service, NOAA, Charleston, SC. 281p.

White, G. F. 1994. Decision or procrastination in floodplain management. Water resources update. Universities Council on Water Resources, No. 97. 4p. http://www.ucowr.siu.edu/updates/pdf/V97_A12.

3

Predicting Trajectories of Urban Growth in the Coastal Southeast

Jeffery S. Allen and Kang Shou Lu

Summary by Jeffery S. Allen and Kang Shou Lu

The growth projected for the Southeast during the next two decades will put enormous pressure on economic, social, and environmental resources. The ability to predict not only how many people will be immigrating to the region, but also the trajectory of growth, is crucial to managing impacts.

In this chapter, we consider models that predict human population growth trajectories at the spatial scales of counties to regions. We focus on hybrid models that integrate a variety of modeling strategies, describing their development and uses. We also describe a model that has been used successfully to predict growth in a tricounty region around Charleston, South Carolina, and discuss the evolution of that model from a three-tiered predictive tool to a neural network-driven protocol.

The growth projection modeling study investigated urban growth in the greater Charleston metropolitan area from 1973 to 1994 and found that over the 21-year period, urban land use growth has exceeded population growth by a 6:1 ratio. The prediction modeling was based on the historical trends of the 1973–1994 study and set under the current policy constraints and the physical environment. For the statistical modeling component of the overall model, a multi-variate logistic regression model was selected because of the nonlinear nature of urban growth problems. A rule-based model was developed to derive the relative transition probabilities of urban growth. This model was designed to complement the pure statistical model primarily through subjective weighting of variables. The third technique used was focus group mapping. A group of experts, local officials, planners, developers, conservationists, and other people were invited to a number of meetings, or interviewed individually, to express their opinions on where growth may occur during the next 30 years. Finally, an integrated GIS model was designed to fully take advantage of the above three models by integrating them into one.

As an additional powerful computational and modeling tool, artificial neural networks have many advantages over conventional mathematical methods and statistical models in addressing complex systems. The use of neural networks for modeling has four major potential benefits: better performance, greater representational flexibility and freedom from current model design constraints, the opportunity to handle explicitly noisy data, and incorporation of spatial dependency, which is currently ignored, in the net representation. The authors applied this model to predict urban growth in coastal South Carolina. The neural network outperformed the conventional logistic regression model in most of the cases (55 of 66) and classification categories (urban, nonurban, overall).

If the current growth trends continue and the predictions hold true, the future urban growth in the modeled areas will sprawl considerably outward from the current urban boundaries. This has several significant economic, environmental, and social implications for policy-making and urban planning. While these implications are too numerous to list here, their importance should not be underestimated and the issues cannot be left unaddressed. It is hoped that modeling projects can help inspire decision-makers and citizens to become involved in the planning processes and land use decisions for areas from the local to regional scales.

3.1 Introduction

The state of South Carolina's population has grown steadily in recent decades, particularly in the coastal counties, the greater Columbia area, and along the Interstate 85 corridor. In the coming 25 years, the population of the state is projected to grow 20 percent, from 3,858,000 to 4,645,000 (U.S. Census Bureau 2001). Over 60 percent of this increase is expected to come from the over-65 population as baby boomers age and retirees move into the state (U.S. Census Bureau 2001). While the bulk of

Georgia's population growth has occurred around the Atlanta metropolitan area, its coastal counties are experiencing population in-migration patterns similar to those occurring in South Carolina.

The rate of conversion of farms and forests to urban uses in South Carolina and Georgia is already significant, and substantially higher than the rate of population growth. According to the U.S. Department of Agriculture's 1999 National Resource Inventory report on the 50 states, Georgia had the nation's third-highest rate of rural acreage conversion to more developed uses between 1992 and 1997, with 1,053,200 rural acres converted, while South Carolina ranked ninth nationally, with 539,700 rural acres converted (London and Hill 2000). For South Carolina, this acreage amounted to a 30.2 percent increase in the amount of developed land in the state over this five-year period. Over this same period, the state's population only grew 5.3 percent.

The growth projected for the Southeast during the next two decades will put enormous pressure on economic, social, and environmental resources. The ability to predict not only how many people will be immigrating to the region, but also the trajectory of growth, is crucial to managing impacts.

In this chapter, we consider models that predict human population growth trajectories at the spatial scales of counties to regions. We focus on hybrid models that integrate a variety of modeling strategies, describing their development and uses. We also describe a model that has been used successfully to predict growth in a tricounty region around Charleston, South Carolina, and discuss the evolution of that model from a three-tiered predictive tool to a neural network-driven protocol.

3.2 Factors Affecting the Prediction of Growth Trajectories

Prediction of the magnitude and trajectory of development, while substantively improved during the past decade by advances in change-detection analysis, geographic information processing, and modeling, remains a difficult task because land use systems and human demographics are inherently complex. Land use systems are multidimensional. Both natural and human factors, including the physical environment, economics, use status, ownerships, and myriad other human activities affect this dimensionality.

Land, either as an element or a subsystem of the natural environment, is the spatial template for numerous processes, including both hydrologic and biogeochemical cycles. The productivity, suitability, capacity, and availability of land are not stable, and changes in these parameters are often unpredictable.

The human part of a land use system is more complex still, and the land use decision-making apparatus reflects this complexity. Decision-making involves government agencies, nonprofit organizations, commercial interests, and the public. These diverse interests endeavor to extract, retain, or protect land values for different purposes. Even the concept of value is multi-faceted (i.e., economic, social, ecological), so there is no single criterion for measuring value. The factors and forces that drive land use change can be either endogenous or exogenous. It is impossible to identify all of these, much less their interrelationships. With such uncertainties, prediction of land use change in an urban system is risk intensive.

Spatially, change in urban land use systems results from new development, infill development, or redevelopment. It may take the form of scattered, "leapfrogging" or concentric spreading from an urban core. It may expand along major roads or diffuse around specific nodes. The difficulty lies in determining its spatial sequence and extent.

The spatial patterns of urban land use are fractal in nature, exhibiting both regularity and irregularity. As a result, the geometry of urban growth can be neither adequately expressed by a single mathematical formula, nor fully described by other physical features in urban space. Urban systems are susceptible to chaotic behavior (Casti 1991) and therefore any change in the urban spatial pattern is predictable only to a limited extent.

3.3 Predicting Growth Trajectories

Despite the difficulty of predicting urban trajectories, a number of comprehensive urban growth models have been developed. Several of these are based on an integrated approach in which a variety of techniques and methodologies are combined to improve predictability and usability for a multiplicity of purposes, stages, and urban scales. TRANUS (de la Barra 1989) and MEPLAN (Echenique et al. 1990) have been used to model residential and employment location, residential and nonresidential floor space, supply and consumption, and goods transport and travel, while taking into account network congestion. Wegener's (1991) Dortmund Model is a spatial model that uses a regional context to simulate intraregional location decisions for industry, residential development, and households, together with associated public policy impacts in the fields of housing and infrastructure. Spatial interaction models and microsimulation methods are used in combination to produce the prediction.

Allen and Lu (2003) took an integrated approach to modeling and predicting urban growth in the Charleston, South Carolina, metropolitan area. A hybrid urban growth model built on a logistic framework was coupled with a rule-based module and a focus group methodology for predicting urban land transitional probabilities (Figure 3.1). Development of the Charleston area (Berkeley, Charleston, and Dorchester Counties) model involved two basic procedures to predict future urban growth in the region. The first was to predict urban transition probabilities with an array of spatial (with geographic coordinates) data. The second procedure was to set urban growth scenarios with aspatial (without geographic coordinates) data. The combination of these procedures yielded maps of future urban growth as well as scenario series (different growth ratios) and temporal series (growth in each successive year) maps. The data used in the Charleston area growth model are listed in Table 3.1.

3.3.1 Charleston Growth Scenario

Two assumptions were involved in the growth scenario. First, it was assumed that the ratio of overall urban land use change (255 percent) to overall population growth (41 percent) over the historic period of 1973 to 1994, a ratio of 6:1, would remain relatively stable for the next 35 years. After much discussion with planners and other interest groups, a more conservative ratio of 5:1 was selected for the final model. Since this ratio is an important index of urban growth, it is used here to determine the spatial scale of the urban development. Second, it was assumed that population would grow as predicted by the Berkeley-Charleston-Dorchester (BCD) Council of Governments with information from the U.S. Census Bureau and the South Carolina State Budget Control Board. In other words, as population increases by 49 percent, from 532,688 persons in 1994 to 795,879 persons in 2030, the total urban area should grow by 245 percent, from 250.07 square miles to 868.55 square miles over the same period. Urban area growth at that ratio is considered

Table 3.1. Charleston-Area Urban Growth Model. Database Construction and Data Preparation.

Data Name	Data Type	Scale Resolution	Spatial Source	Date	Data
Source data.					
Nature Environ					
Digital Elevation	Grids	24,000	30 m	Variable	SCDNR
Hydro/Water	TIGER/Line File	100,000		1990	Census Bureau
Soils					
Land Use					
Urban Land Use	Images		30–69 m	1973–94	BCD COG
Land Use76	GIRAS	100,000	90 m	1976	USGS
Land Use89	Images (Classified)		60 m	1989	USC
Incorporated Areas TIGER/Line File	100,000			1990	US Census 90
Expert predicted Population/Housing	Polygon			1998	SCCCL
Population	TIGER/Line File	Census Block		1990	Census Bureau

Table 3.1. Charleston-Area Urban Growth Model. Database Construction and Data Preparation. (*Continued*)

	Table	County		
Projected Pop	TIGER/Line File	Census Block	1999	BCD COG
HU Mean Value			1990	Census Bureau
Infrastructure/Facilities				
Roads	TIGER/Line File	1:100,000	1997	Census Bureau
Improved Roads	Line	1:100,000	1998	BCD COG
Planned Roads	Line	1:100,000	1999	BCD COG
Planned Bridges	Line		1999	BCD COG
Water Line	Line		1994	Dept of Com- merce
Sewer Line	Line		1994	Dept of Com- merce
Approved W	Points		1999	BCD COG (Not Available)
Policy/Social				
Forest Boundaries	Polygon	24,000		SCDNR
Wetland	Polygon	24,000	1989	SCDNR (NWI)
Protected Areas	Polygon	24,000	1983	SCCCL
Zoning Districts	Polygon	Not available		

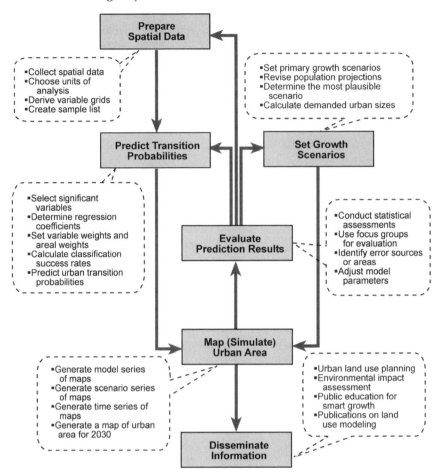

Fig. 3.1. Framework of the integrative model for urban growth in the Charleston, South Carolina, region

to be conservative because first, it is widely perceived that overall population growth in the Charleston area is underestimated, and second, a higher ratio of urban land use increase per unit population growth has been observed in nearby Mt. Pleasant and Summerville since 1994. National growth rates for population and urban area expansion support this notion. According to Rusk (1997), many metropolitan areas such as Detroit (13:1), St. Louis (7:1), and Baltimore (5:1), not necessarily targeted as high-growth areas, have seen similar or higher ratios between 1960 and 1990 over a longer term than the current study used.

3.3.2 Transition Probabilities

For the prediction of urban transition probabilities, four techniques, including logistic regression possibility modeling, rule-based modeling, focus group mapping, and integrated GIS modeling, were used in the project. Because the size of the region is too large for high-resolution modeling (parcel level), analysis units were set to 250X250 meters. All the source data were resampled at this resolution before further processing, then remapped at 30X30 meter resolution for final map production. Some information losses associated with this resampling were expected.

3.3.3 Logistic Regression Model

For the statistical modeling, a multivariate logistic regression model was selected because of the nonlinear nature of urban growth problems. Urban growth was measured only in terms of change in urban area or urban land use. Urban land use is the binary dependent variable, while independent variables are a mix of continuous, discrete, and dichotomous variables that represent the major physical, economic, and social factors influencing urban growth and land use. Fifteen variables were used to build the model. The selection of sets of variables was limited by the availability of data. All variable grids were resampled with a cell size of 250 meters to fit the analytical capabilities of the software and hardware. The model was calibrated using the 1989 data set and validated using the 1994 data set, the most recent urban land use data available. Only those variables that proved to be statistically significant were used to rebuild the prediction model. All predictions of future growth were based on the new input and the 1994 conditions. The resulting map from the logistic regression is shown in Figure 3.2.

3.3.4 Rule-Based Model

A rule-based model was developed to further enhance the relative transition probabilities of urban growth. This model was designed to complement the pure statistical model for three reasons. First, the overall success rate of the statistical model is relatively low for land use change prediction though the model is statistically significant. This is particularly true if urban and suburban areas are spatially scattered or

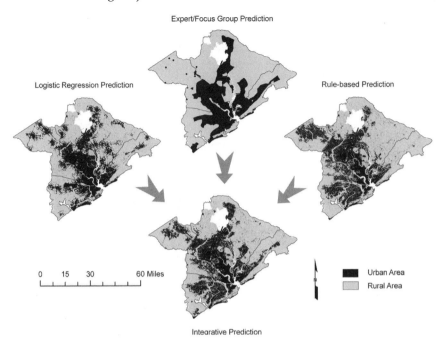

Fig. 3.2. The integrative urban growth model for the Charleston, South Carolina, region

fragmented, as is the case in the Charleston region. Second, some statistical predictions do not make any sense spatially. This is especially true when using a model built on historical data over a short period but predicting the future growth over a relatively long period. Third, the statistical model used here is not sensitive to some policy variables that are actually very important factors affecting urban growth. Nor is this model sensitive to new or planned primary roads that greatly influence where new development will occur.

The rule-based model utilizes various rating, weighting, and ranking techniques through map overlays and map manipulation (map algebra) to create a map of ranked suitabilities or relative probabilities of urban transformation. The former is a common practice in geographic information processing. Although this method is conditional on subjective decisions made by the modelers, the rules used for assigning different values, selecting rating scales, or determining weighting factors are all based on theories and knowledge of urban growth and planning. In this prediction, protected land, wetlands, and water were excluded by assigning the value 0 to them, while other variables, such as proximity to improved primary roads, distance to water lines, and adjacency to commercial centers, were weighted higher during the overlay process. The map that resulted from the rule-based model is shown in Figure 3.2.

3.3.5 Focus Group Mapping

The third technique used in the growth modeling process was focus group mapping. A group of experts, including local public officials, planners, developers, and conservationists who possess a profound knowledge of the area and of urban growth issues was invited to a number of meetings, or interviewed individually to express their opinions on where they think the urban area is going to expand during the next 35 years. A base map of the BCD region was provided and each member of the focus group was asked to mark on the map, or describe in writing, the urban boundary for year 2030. These urban boundaries were further discussed during a second focus group meeting and a generally agreed upon urban boundary was drawn. Water, wetland, and protected land were excluded for this study in order to calculate the net growth area. Although the projected urban area was not differentiated in terms of priority or probabilities, the method provides a unique insight into perceived future growth patterns. No growth scenarios were involved in this step of the model. The map that resulted from the focus group prediction is shown in Figure 3.2.

3.3.6 Integrated GIS Model

An integrated GIS-based model is designed to take advantage of the above three models by integrating them into one. In this model, the focus group prediction was weighted 10 percent while the other two predictions were weighted 45 percent each. This was done to eliminate the arbitrary boundary of the focus group prediction but keep the spatial differentiation of transition probabilities predicted by the logistic and rule-based models. The latter allows modelers to map the spatial extents of future urban growth based on different growth scenarios as set previously. The model is intended to keep the insight of urban growth predicted by the focus group members; to integrate the policy factors with physical and economic factors in the rule-based model; and to keep the detail and objectivity of the logistic regression model. The map that resulted from the integrated GIS prediction is shown in Figure 3.2. Results of the integrated model also allow us to predict the relative increase in urban area through a time series map (Figure 3.3).

3.3.7 Implications

If current growth trends in the Charleston metropolitan area continue and the predictions hold true, future urban growth will exhibit a pattern typical of urban sprawl. Economically, there will be increased pressure on urban infrastructure and a decrease in the efficiency of natural and urban resource use. As the urban area sprawls, population density and housing density will decrease. The spreading of residential, commercial, industrial, and service centers will require additional intraurban roads, bridges, water and sewer lines, as well as other services and infrastructure, such as fire stations, schools, and medical facilities. If the allocation of urban resources is per capita based, the overall travel time from home to work, to shopping centers, to schools,

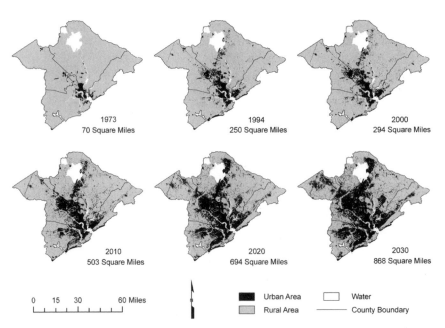

Fig. 3.3. Historical and future urban growth in the Charleston, South Carolina, region, 1973–2030

and to other service areas will increase substantially as these facilities and services are distributed over a wider area. Even if the allocation of urban resources is area based, the increase in per capita urban land use, infrastructure, and services will decrease the efficiencies-limited resource use. How to balance this need for urban growth with the efficient use of resources is a daunting issue for policy-makers and urban planners.

Urban sprawl implies the large scale conversion of natural and rural land to urban use. This conversion often involves damaging, destroying, or altering natural environments by physically removing the elements of ecosystems, building barriers to natural processes, disposing pollutants that may alter natural geochemical processes and cycles, and so on. Even if policies and regulations are implemented to protect some areas from being developed, they cannot guarantee that these protected areas will not be polluted, for example by urban run-off generated upstream. How to make an urban plan that will protect the wetlands, endangered species, and unique landscapes and riparian resources is a challenge to planners and citizens.

3.4 Neural Networks and the Future of Growth Trajectory Forecasting

Neural network computing and generic algorithms have been proposed as alternative approaches to handling complex systems (Sui 1997). Although the first neural network models appeared as early as the 1940s (McCulloch and Pitts 1943), they were applied principally in signal processing (Widrow and Stearns 1985), control (Nguyen and Widrow 1989; Miller et al. 1990), pattern recognition (Le Gun et al. 1990), medicine (Anderson 1986; Anderson et al. 1986; Hecht-Nilsen 1990), speech production and recognition (Sejinowski and Rosenberg 1986; Lippmann 1989), and business (Collins et al. 1988). Following a period of active neural network development in the 1950s and 1960s, interest in the approach waned during the 1970s. Robust propagation training methods and the advancement of computer technology led to renewed interest in neural networks in the 1980s and early 1990s (Fausett, 1994). Not until the mid-1990s, however, were neural networks used for land use modeling and resource management.

Openshaw (1993) used a neural network for modeling spatial interaction. Wang (1994) used artificial neural networks in a geographic information system for agricultural land suitability assessment. Gimblett et al. (1994) developed a forest management decision model, based on neural networks and generic algorithms, that was validated in the Hoosier National Forest. Fischer and Gopal (1994) used an artificial neural network approach to model interregional telecommunications flows. Gong (1996) combined evidential reasoning and artificial neural network techniques for geological mapping. More recently, Yeh and Li (2002) used neural networks and cellular automata to simulate potential or alternative urban development patterns based on different planning objectives.

As a powerful computational tool, artificial neural networks have several advantages over conventional mathematical methods and statistical models in addressing complex systems. According to Openshaw and Openshaw (1997), the use of neural networks for modeling has four major potential benefits: better performance, greater representational flexibility and freedom from current model design constraints, the opportunity to handle explicitly noisy data, and incorporation of spatial dependency, which is currently ignored, in the net representation.

According to Kohonen (1988), neural networks are "massively parallel interconnected networks of simple (usually adaptive) elements and their hierarchical organizations which are intended to interact with other objects of the real world in the same way as the biological nervous systems do." Biological nervous systems are composed of millions of interconnected neurons. Each neuron (Figure 3.4) has three functioning parts: dendrite, soma, and axon. Dendrites are sensors or signal receivers. Soma are the information processing centers, and the axons are the sender or signal emitters. There is a synaptic gap between the axon and the dendrite of two adjacent neurons. When a neuron receives signals from other neurons through its dendrites, the soma starts processing these signals, assigning each a different weight, and making a summation of all weighted signals. Then, the soma applies a signal transfer function or activation function to calculate the overall effect of these weighted signals. If the overall effect is strong enough, the neuron will become activated and send its own reactive signal through the axon to another neuron. When millions of biological neurons are networked, the structure and information transfer processes are far more complicated, but the architectural and functioning principles can be used for constructing artificial neural networks.

Artificial neural networks differ mainly in architecture, activation functions, and methods for determining weights and biases. Fischer

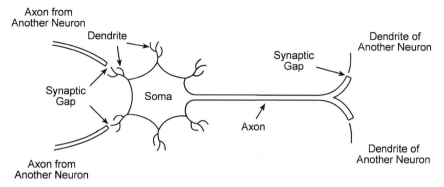

Fig. 3.4. Biological neuron

(2001) has identified four types of neural network models that seem particularly attractive for spatial analysis. They include back-propagation networks, radial basis function networks, adaptive resonance theory (ART) based models (supervised and unsupervised), and self-organizing feature map networks. For urban land use prediction, the three-layer, sigmoid function-based, back-propagation neural networks appear to be most appropriate.

We developed a predictive land use model based on back-propagation neural networks (Figure 3.5). In this model, artificial neurons, called units or nodes, are arranged in three layers: one input layer, one hidden layer, and one target layer. The input layer has multiple units representing predictor variables, but the target layer has only one unit representing a dichotomous variable with two values: 0 for undeveloped or rural and 1 for developed or urban. The hidden layer can have multiple units, but not more than the input units and not fewer than the target units. Units between two adjacent layers are interconnected, which creates a hierarchical architecture. The sigmoid function has a value range from 0 to 1, appropriate for handling binary classification problems. Weights and biases are learned from samples using a robust algorithm through a feedback error propagation process.

Training such a network involves three phases: feed-forward prediction, backward error propagation, and adjustment of biases and weights. During the phase of feed-forward prediction, each hidden unit receives the signal (value) from each input unit, sums all the received signals with different weights and biases, and applies a sigmoid function to the sum and sends the output to each target unit. The target unit receives a signal (value) from each unit in the hidden

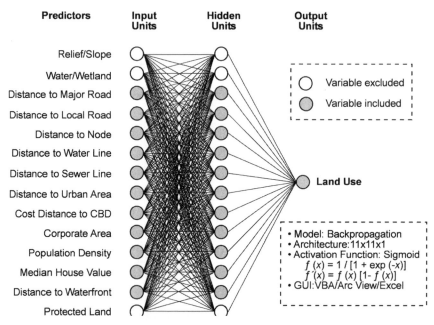

Fig. 3.5. The neural network model for land use change in the Myrtle Beach, South Carolina, region

layer, sums all signals (values) with corresponding weights and biases, and applies the sigmoid function. The output of the target layer is the prediction of the neural network. During the error back-propagation phase, the network compares the predicted value with the observed value to calculate the error and the total error correction term using a gradient reduction method. The total error correction term is then distributed over each hidden unit based on its weight and bias. At the same time, the weight and bias correction terms are calculated respectively for each hidden unit. The same process is repeated for each input unit but only the weight and bias are calculated. During the last phase, all weights and biases are updated using their respective correction terms obtained in the previous phase. The process is repeated for every training pattern (record) from the data set over a complete cycle called an epoch. The training process continues iteratively until a stop condition is met. The ultimate goal is to minimize the error and build a network that is sufficiently trained with the ability to approximate a complex land use system but not overfitted (which would limit future prediction power).

We applied this model to predict urban growth in coastal South Carolina. Three different study areas were selected: the Myrtle Beach Region (Allen and Lu 2004, unpublished manuscript), Beaufort County (Lu, Allen, and Liu 2005), and Hilton Head Island (Lin et al. 2005) to test the reliability and validity of the neural net models for predicting land use change at regional, county, and local scales, respectively. The neural network outperformed the conventional logistic regression model in most cases (55 of 66) and classification categories (urban, nonurban, overall). Statistically, the network improved the urban classification accuracy by an average 10.74 percentage points in the Myrtle Beach region. Spatially, the neural network was able to differentiate dispersed urban land parcels from predominantly undeveloped rural environments where the logistic regression often failed to generate reliable predictions (Figure 3.5). Risks associated with prediction errors were also much smaller than those for the logistic regression model when different classification strategies were used.

Each geographic unit is unique and no two urban areas are the same. While we will always need an empirical approach to study land use and urban systems, neural networks are one of the more promising alternative approaches and methodologies for modeling and predicting their changes. As database management, computing technologies, and modeling techniques continue to evolve, prediction of the trajectory of development will become more accurate. However, the chaotic aspects of natural, social, and economic forces will always challenge the predictive process.

3.5 Discussion

Population growth is still the most fundamental driving force that affects land use change and urban growth. Understanding the future trajectory of population growth and urban development is critical for advanced land use planning and adaptive resource management. Although land use systems are constantly changing, their environmental impacts are not always positive. All too often, detrimental changes in coastal ecosystems are not perceived detrimental unless their negative impacts have cumulatively reached certain critical levels that are too severe to ignore, too late to avoid, or too costly to mitigate. Predictive modeling does not necessarily enable us to foresee a brighter future, but it does provide prediagnosed warning signals for some of the potentially unexpected consequences of development patterns.

Public policy makers are often reluctant to advocate environmentally friendly growth regulations when they are concerned that such regulations may have economic costs or limit economic development. Likewise, land developers are seldom eager voluntarily to save species or habitats if additional land management expenses are expected. Without quantified descriptions and specific locations of potential impacts, both planners and natural resource managers become less persuasive and effective in promoting ecologically sustainable planning and development. Predictive modeling is not only a necessary step toward anticipating the potential ecological effects of land use change (Pearson et al. 1998), but it is also an important mechanism for generating quantitative, spatial, temporal, and visual information crucial for urban planning, impact assessment, and public education.

Acknowledgment

Support for the work reported have was provided in part by the Land Use-Coastal Ecosystem Study under NOAA Coastal Ocean Program Grant NA960PO113.

References

Allen, J. and K. Lu. 2003. Modeling and prediction of future urban growth in the Charleston region of South Carolina: A GIS-based integrated approach. *Conservation Ecology* 8(2):2.

Allen, J. and K. Lu. 2004. Artificial neural network vs. binary logistic regression: Two alternative models for predicting urban growth in the Myrtle Beach region. Report submitted to NOAA's Land Use-Coastal Ecosystem Study (LU-CES) Program. (Unpublished manuscript, 26p.)

Anderson, J. A. 1986. Cognitive capabilities of a parallel system. In, Bienestoch, E., F. Fogelman-Souli, and Weisbuch (eds.), Disordered Systems and Biological Organization. NATO ASI Series, F20, Springer-Verlag, Berlin.

Anderson, J. A. and E. Rosenfeld, editors. 1988. Neurocomputing: Foundations of research. MIT-Press, Cambridge, Mass.

Anderson, J. A., R. M. Golder, and G. L. Murphy. 1986. Concepts in distributed systems, pp. 634:260–272. In, Szu, H. H. (ed.), Optical and Hybrid Computing. Society of Photo-Optical Instrumentation Engineers, Bellington, WA.

Casti, J. L. 1991. Searching for Certainty: What Scientists Can Know about the Future. William Morrow, New York, NY. 496p.

Collins, E., S. Ghosh, and C. L. Scofield. 1988. An application of multiple neural network learning system to emulation of mortgage underwriting judgments. *IEEE International Conference on Neural Networks* 11:459–466. San Diego, CA.

de la Barra, T. 1989. Integrated Land Use and Transport Modeling. Cambridge University Press, Cambridge, UK. 179p.

Echenique, M. H., A. D. Flowerdew, J. D. Hunt, T. R. Mayo, I. J. Skidmore, and D. C. Simmonds. 1990. The MEPLAN models of Bilbao, Leeds and Dortmund. *Transportation Reviews* 10:309–322.

Fausett, L. 1994. Fundamentals of Neural Networks: Architectures, Algorithms, and Applications. Prentice-Hall, Englewood Cliffs, NJ. 461p.

Fischer, M. M. 2001. Computational neural networks — Tools for spatial data analysis, pp. 15–34. In, Fischer, M. M. and Y. Leung (eds.), Geo-Computational Modeling: Techniques and Applications. Springer, New York, NY.

Fischer, M. M. and S. Gopal. 1994. Artificial neural networks. A new approach to modeling interregional telecommunication flows. *Journal of Regional Science* 34:503–27.

Gimblett, R. H., G. L. Ball, and A. W. Guisse. 1994. Autonomous rule generation and assessment for complex spatial modeling. *Landscape and Urban Planning* 30:13–16.

Gong, P. 1996. Integrated analysis of spatial data from multiple sources: Using evidential reasoning and artificial neural network techniques for geological mapping. *Photogrammetric Engineering and Remote Sensing* 62:513–523.

Hecht-Nielsen, R. 1990. Neurocomputing. Addison-Wesley, Reading, MA. 433p.

Kohonen, T. 1988. The "Neural" phonetic typewriter. *Computer* 21(3):11–22.

Le Gun, Y., B. Boser, J. S. Denker, D. Henderson, R. E. Howard, W. Hubbard, and L. D. Jackel. 1990. Handwritten digit recognition with a backpropagation network, pp. 396–404. In, Touretsky, D. S. (ed.), Advances in Neural Information Processing Systems 2. Morgan Kaufman, San Mateo, CA.

Lin, H., K. S. Lu, M. Espey, and J. S. Allen. 2005. Modeling urban sprawl and land use change in a coastal area: A neural network approach. Paper presented at the American Agricultural Economics Association Annual Meeting, Providence, Rhode Island. 20p. http://agecon.lib. umn.edu/cgi-bin/pdf_view.pl?paperid = 16329&ftype = .pdf.

Lippmann, R. P. 1989. Review of neural networks for speech recognition. *Neural Computation* 1:1–38.

London, J. B. and N. L. Hill. 2000. Land conversion in South Carolina: State makes the Top 10 list, pp. 2–3. Jim Self Center on the Future, Strom Thurmond Institute, Clemson University, Clemson, SC. http://www.strom.clemson.edu/

Lu, K. S., J. S. Allen, and G. Liu. 2005. Using innovative neural networks to simulate and assess land use change and its ecological impact under different growth scenarios. Paper submitted to the Inter-disciplinary Environmental Association for the 11th International Conference on the Environment, Orlando, Florida. 25p.

McCulloch, W. S. and W. Pitts. 1943. A logical calculus of the ideas immanent in nervous activity. *Bulletin of Mathematical Biophysics* 5:115–133.

Miller, W. T., R. S. Sutton, and P. J. Werbos (eds.). 1990. Neural Networks for Control. MIT Press, Cambridge, MA. 544p.

Nguyen, D. and B. Widrow. 1989. Fast learning in networks of locally tuned processing units. *Neural Computation* 1:281–294.

Openshaw, S. 1993. Modeling spatial interaction using a neural net, pp. 147–164. In, Fischer, M. M. and P. Nijkam, (eds.), GIS Spatial Modeling and Policy. Springer, Berlin.

Openshaw, S. and C. Openshaw. 1997. Artificial Intelligence in Geography. John Wiley and Sons, Hoboken, NJ. 348p.

Pearson, S. M., M. G. Turner, and J. B. Drake. 1998. Landscape change and habitat availability in the Southern Appalachian Highlands and Olympic Peninsula. Journal of Ecological Applications, 9(4):1288–1304.

Rusk, D., J. Blair, and E. D. Kelly. 1997. Debate on the theories of David Rusk. Edited transcript of proceedings in *The Regionalist* 2(3):11–29.

Sejnowski, T. J. and C. R. Rosenberg. 1986. Nettalk: A parallel network that learns to read aloud. The Johns Hopkins University Electrical Engineering and Computer Science Technical Report JHU/EECS-86/01, 32. Reprinted in Anderson and Rosenfeld, pp. 663–672.

Sui, D. Z. 1997. The syntax and semantics of urban modeling: Versions vs. visions. Paper submitted to the Land Use Modeling Workshop at U.S. Geological Survey EROS Data Center, Sioux Falls, SD. 17p.

U.S. Census Bureau. 2001. Population and housing unit counts: Population from 1790 to 1990. http://www.Census.gov/population/censusdata/table 16.pdf.

Wang, F. 1994. The use of artificial neural networks in a geographical information system for agricultural land-suitability assessment. *Environment and Planning A* 26:265–284.

Wegener, M., R. L. Mackett, and D. C. Simmonds. 1991. One city, three models, comparisons of land use/transport policy simulation models for Dortmund. *Transport Review* 11:107–129.

Widrow, B. and S. D. Stearns. 1985. Adaptive Signal Processing. Prentice-Hall, Englewood Cliffs, NJ. 528p.

Wu, F. and C. J. Webster. 1998. Simulation of land development through the integration of cellular automat and multi-criteria evaluation. Environment and Planning B. *Planning and Design* 25:103–126.

Yeh, A. G. O. and X. Li. 2002. Urban simulation using neural networks and cellular automata for land use planning, pp. 452–464. In, Richardson, D. and P. Van Oosterom, (eds.), Advances in Spatial Data Handling. Springer, Berlin.

4

Urban Typology and Estuarine Biodiversity in Rapidly Developing Coastal Watersheds

G. S. Kleppel, Dwayne E. Porter, and M. Richard DeVoe

Summary by Susan E. Ferris

Modern suburbs, those built after World War II, tend to be disconnected from the "urban core," with considerable infrastructure and associated impervious surfaces (e.g., roads and parking lots) being required to accommodate residents and traffic. Modern suburbs depend heavily on the use of automobiles for transportation because single-purpose zoning separates residential from commercial and public areas by distances not easily traversed on foot, and the distribution of population is often not conducive to public transportation.

By contrast, traditional urban developments—those built prior to 1945—such as small towns, villages, and cities tend to combine residential and commercial zones into mixed-use districts that are compact and pedestrian friendly. Historically, walkable access to urban services was a requirement of the urban environment. As such, traditional communities tend to have fewer, often narrower, roads, than do suburbs and, for a variety of reasons (including a revulsion of insect-infested marshes and flood prone waterways), greater riparian buffering than suburbs.

A key difference between suburban and traditional urban land use relates to population density. Due in large part to the infrastructure requirements of modern zoning ordinances, subdivision and commercial development codes, and, at least early in the development of a suburb, the large parcels provided for single-family residences, typical suburban population densities tend to be low, often less than one person per acre. For instance, in the southeastern United States, between three and eight acres of urbanized land are required to support each person in the community. Conversely, in traditional small towns and many of the cities developed prior to World War II that are today considered as high-quality development (e.g., Charleston, South Carolina, or Savannah, Georgia), population densities between 14 and 20 people per acre are typical. That is, there is far less than an acre of urban space per person. Apparently, population density and quality of life have little to do with one another.

The southeastern United States is currently undergoing massive population growth, particularly in environmentally sensitive coastal areas. Rural areas, consisting primarily of agricultural communities and small towns, are being replaced by suburban subdivisions. Current research suggests that the new development is contributing to contaminant loading derived from the runoff of roads, rooftops, driveways, sidewalks, and parking lots. The natural capacity of ecosystems to process contaminants is hampered by the infrastructure associated with sprawling development. Documented consequences of coastal watershed urbanization include a decrease in species diversity, increased eutrophication or excess nutrients from fertilizers and road runoff, disruption of aquatic plant communities, a decreased ability of the ecosystem to process excess nutrients, and reduced numbers of invertebrate (i.e., shellfish) and vertebrate (i.e., fish) stocks.

Simple models can be used to make a rough comparison of the impacts of two different styles, or typologies, of development, suburbs and traditional urban communities, on coastal ecosystems. Such a comparison is based in part on the species-area relationship (SAR), which predicts that the number of species (or species richness) that a landscape supports changes systematically with the area of that landscape. One can ask if development influences the SAR. Using a data set from Charleston County, South Carolina, collected by the South Carolina Department of Natural Resources during the Tidal Creeks Project (1996), it was shown that as the area of a watershed (the small drainage bounded by ridgelines, drained by tidal creeks) increases, the number of species of small, bottom-dwelling invertebrates living in the tidal creeks in the watershed also increases. However, if that watershed is developed, the tidal creek will support only about half as many species per unit area as an undeveloped creek. Potentially, as more watersheds in a region are developed, the number of species in the tidal creeks of an "average watershed" will decline. Therefore, development styles that are land intensive (require a great deal of urban land per person), such as suburbs, will use up more of the watershed area in a region and should experience a greater decline in the average species richness in the area's watersheds than development styles that require less land per person, such as traditional urban communities.

Using U.S. Census Bureau estimates of expected population growth in the eight coastal counties of South Carolina, along with data available in a geographic information system on the sorts of land covers and uses in the counties, it was possible to estimate the watershed area in the coastal counties and the number of people who would be moving to these watersheds between 1995 and 2025. One can then ask how much space in the watershed the in-migrant population would require if the watersheds were developed as suburbs (conservatively requiring 3 acres/person) or as traditional communities (conservatively requiring 1 acre/person). Finally, one can roughly evaluate how the increase in developed watershed area would impact invertebrate species richness in the tidal creeks associated with these watersheds.

What emerges from this simple exercise is that we find that virtually the entire estimated watershed areas of three counties (Horry, Berkeley, and Beaufort) will be developed if suburbs are the primary typology used to accommodate future growth. With a modest change in density, to one urban acre per person, however, none of the coastal counties will experience complete watershed development. Furthermore, using conservative suburban density estimates, invertebrate species richness will decline by an average of 24 percent in South Carolina's tidal creeks over the period from 1995 to 2025. The decline will be 38 percent, on average, in the three most rapidly developing counties. With even the modest change in density used here (from 3 to 1 acre/person), the average loss of tidal creek species is reduced to 9 percent coastwide and 16 percent in the three fastest-growing coastal counties.

Although these predictions are coarse, they suggest that alternative urban typologies have different environmental impacts. At truly traditional population densities of 20 people per acre, the impact of development on tidal creek species richness would predictably be even smaller than those estimated by densities of one person per acre. The point is that low-impact alternatives to suburban development in the coastal zone appear to exist in the form of traditional small town design. Studies indicate that these traditional alternatives often perform well in the market.

4.1 Introduction

Although it is recognized that urbanization will be among the leading causes of ecosystem disruption during the twenty-first century (Vitousek 1994; Pimm and Raven 2000; Sala et al. 2000), it is unlikely that the rate of conversion of natural and agricultural landscapes to urban use will abate any time soon. The impacts of urban encroachment on ecosystems will depend in part on the spatial extent over which development occurs. The term "urban" has been defined in a variety of ways by different authors. In this paper, we consider an urban environment as a relative level of human social and structural development on the landscape according to the definition of Kleppel et al. (2004). Urban environments exist where human population densities exceed the average for the surrounding landscape and where infrastructure and urban services exceed those found on the surrounding landscape. Thus, urban environments may include small towns and villages, and modern suburbs on rural and urban fringe landscapes, as well as small and large cities and groups of cities. Although all forms of urban development tend to reduce the functionality of natural ecosystems (Kleppel 2002), it is worth asking whether certain kinds of development have lower impacts than others. We are broadly concerned with the question of whether the spatial distributions of urban attributes, i.e., urban typology, influence the kinds and magnitudes of impacts to associated ecosystems (Lowrance 1998; Wenger 1999; Kleppel 2002; Kleppel et al. 2004).

Certain urban typologies are associated with specific periods in the history of American landscape architecture. A benchmark in landscape architecture occurred with the emergence of the modern suburb, shortly after World War II, circa 1947. Modern suburban typologies differ markedly from their pre-World War II counterparts, which are connected, as fringing neighborhoods, to the existing urban core. After World War II, suburbs began to disconnect from the urban core and to alter key spatial relationships associated with human mobility and the infrastructure needed to support that mobility (Katz 1994; Eisenberg 1998). The modern suburb, with its characteristic single-purpose zones, has redefined the American mobility scale to that of the automobile, and in doing so, has necessitated the commitment of extensive portions of the landscape to (largely impervious) supportive infrastructure (Duany et al. 2000). Alternatively, traditional urban typologies represented by small towns, villages, and cities developed

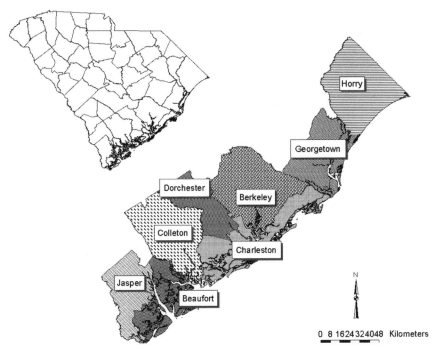

Fig. 4.1. South Carolina county boundaries. The eight coastal counties are shaded and identified by name

prior to 1950 (see Kleppel et al. this volume) are spatially constrained, and defined by large riparian buffers and reduced impervious surface area (Kleppel 2002; Kleppel et al. 2004), all features that are today considered characteristic of "smart" or "green" design (Arendt 1992; Duany et al. 2000).

The modern suburb is currently the dominant urban typology in the southeastern United States, where one of the nation's highest regional rates of urbanization is underway (Culliton et al. 1990; Porter et al. 1996; Kleppel and DeVoe 2000; U.S. Census Bureau 2000). Significant portions of coastal ecosystems have already been lost to development (Porter et al. 1996), with no hope of replacing the critical functions they once fulfilled. Ecosystem structure and function, altered by physical disruption and the associated loss of capacity to process the increased point and nonpoint contaminant loads that urban landscapes produce, are quickly degraded (Scheuler 1995; Collins et al. 2000). Declines in species diversity (Lerberg 1997), increased eutrophication (Valiela and

Teal 1979; Paerl 1988, 1997; Nixon 1995), disruption of macrophytic community structure (Pennings and Bertness 1999), alteration of ecosystem metabolism (Cai and Pomeroy 1999), and reduced recruitment to invertebrate and vertebrate fisheries (Nixon 1980; Hoss and Engel 1996) are among the documented consequences of urbanization of coastal watersheds.

In this paper, we use data from a study of tidal creeks in Charleston County, South Carolina (Holland et al. 1996), along with geographic and demographic data from South Carolina's coastal counties (Figure 4.1) to investigate the impacts on tidal creek benthic macroinvertebrate biodiversity of suburban and traditional-urban typologies in the watersheds that they drain. We define a watershed as a small drainage on the coastal plain that is spatially constrained by ridgelines, and we consider the null hypothesis that differences in tidal creek benthic biodiversity are unrelated to urban typology in the watershed.

4.2 Methods

4.2.1 Data

Benthic macroinvertebrate samples were collected by the South Carolina Department of Natural Resources (SC DNR) in 24 tidal creeks in Charleston County, South Carolina, between July 5 and September 10, 1994, as part of the Charleston Harbor Tidal Creeks Study (Holland et al. 1996). Samples were taken with a 3.8 cm internal diameter PVC corer (area = 45.6 cm^2) to a depth of 15 cm. The number of samples collected in each tidal creek was a function of creek length. Sampling and sample processing are described in detail by Lerberg (1997) and Lerberg et al. (2000).

The tidal creeks were located in small drainages, or watersheds (13.1 to 442.9 ha), defined by topographic ridgelines (Table 4.1). Lerberg et al. (2000) distinguished several land use and cover classes in the upland (developable) portions of these watersheds, namely: (1) forested, (2) agricultural, (3) suburban, (4) urban, and (5) industrial. We combined classes to distinguish watersheds as either "undeveloped" or "urbanized." Undeveloped watersheds were forested (class 1). Urbanized watersheds contained one or more of the land uses in classes 3 to 5 (suburban, urban, industrial). We did not include the single watershed considered agricultural. We also omitted Horlbeck and Long Creeks, which Lerberg considered undeveloped (i.e., upland forest) but we considered mixed use. While sparsely populated, these watersheds contained >20 ha (10–25% of the watershed) of land in agricultural use. Long Creek also contained 6 ha in urban use. Our data set consisted of 6 creeks in undeveloped watersheds and 11 creeks in urbanized watersheds. The mean watershed area, A_M, was 126.3 ha.

4.2.2 Theoretical Context

It has been known for years that as the area of a landscape increases, the number of species on that landscape also increases in a roughly exponential way (Arrenhius 1921; Preston 1962). That correlation has

Table 4.1. Characterization of Tidal Creeks Sampled Near Charleston, South Carolina (cf. Lerberg 1997 and Lerberg et al. 2000).

Creek Name	Watershed Class (Lerberg et al. 2000)	Watershed Class (this study)	Watershed Area (ha)
Deep Creek	Forested	Undeveloped	64.7
Foster Creek	Forested	Undeveloped	16.0
Lighthouse Inlet Creek	Forested	Undeveloped	37.3
Rathall Creek	Forested	Undeveloped	72.1
Lachicotte Creek	Forested	Undeveloped	13.1
Beresford Creek	Forested	Undeveloped	25.1
Diesel Creek	Industrial	Urbanized	105.3
Koppers Creek	Industrial	Urbanized	116.6
Shipyard Creek	Industrial	Urbanized	279.1
Vardell Creek	Urban	Urbanized	68.7
New Market creek	Urban	Urbanized	187.2
Bull Creek	Suburban	Urbanized	442.8
Metcalfs Creek	Suburban	Urbanized	129.7
Shem Creek	Suburban	Urbanized	427.8
Yatch Club Creek	Suburban	Urbanized	69.1
Cross Creek	Suburban	Urbanized	312.9
Parrot Creek	Suburban	Urbanized	147.2

been called the species-area relationship, or SAR. Recent studies reveal that what one considers the relevant landscape can influence the SAR significantly (Plotkin et al. 2000; Crawley and Harral 2001). In the present study, we consider that the species richness, S, of tidal creek benthic communities varies as a function of the area, A, of the associated watershed, according to a power law (MacArthur and Wilson 1967),

$$S = cA^z, \tag{4.1}$$

where c is a scaling factor. The slope, z, is related to habitat connectivity and complexity (Forman and Godron 1996) and is influenced by numerous landscape variables (Plotkin et al. 2000; Crawley and Harral 2001). Since, in the present study, the number of samples, n, varied with creek length, species density, S′, defined as S/n, was substituted for S as a measure of biodiversity.

4.2.3 Demographic and Land Cover Data

We used the SAR, along with a coarse analysis of population growth rates and development trajectories, to predict the impacts of various urban typologies on benthic biodiversity in the tidal creeks of South Carolina's eight coastal counties. We assumed that at the relatively coarse level of analysis used in these projections, tidal creeks along the entire South Carolina coast are comparable. Demographic projections by county for the year 2025 were obtained along with 1995 population estimates from the U.S. Census Bureau (2000). Land cover estimates were derived from Thematic Mapper imagery classified by the South Carolina Department of Natural Resources and transferred to the University of South Carolina's Geographic Information Processing Laboratory. The 1995 composite land cover data for South Carolina's coastal counties were processed with ArcView® software and reformatted for use in this analysis (Table 4.2). The land area in each county that was considered "unavailable for development" was estimated as the sum of the areas of wetland, open water, and existing urban development (areas that were classified as urban in 1995 were considered unavailable for further development). We assumed that the total watershed area in a given county is equal to the sum of open water and wetlands areas.

The change in average benthic species density in the tidal creeks in a given county that was due to a particular urban typology (suburban, traditional) was estimated by first determining the amount of

Table 4.2. Land Cover and Designations of Developable and Undevelopable Land in the Eight Coastal Counties of South Carolina.

	Horry	George-town	Charleston	Berkeley	Dorchester	Colleton	Beaufort	Jasper
Total county area (ha)	323647	253593	318743	351021	148282	288090	237118	175963
water/wetlands (ha)	130869	113469	193716	78412	19983	78040	157418	44067
existing urban (ha)	56305	46409	33037	81703	8610	31955	64002	18197
total unavailable (ha)	74902	56398	96182	60601	51504	84837	32126	53139
total developable (ha)	185081	139358	237662	149743	127264	209630	79383	131304
developable (% of total)	57	55	75	43	86	73	33	75
total watershed area (ha)	130869	113469	193716	78412	19983	78040	157418	44067

land in a watershed of average area, A_M, required by that typology to support the expected human population growth (see below). We then computed S' for the watershed by summing the species densities in the developed and undeveloped portions of the watershed.

An estimate of the "new" population arriving in coastal county watersheds between 1995 and 2025 was computed from county population estimates, assuming that the proportion of the population in the watersheds reflects the proportion of watershed area to the total developable area of the county. Thus,

$$N_W = N_C \times A_W/A_C, \qquad (4.2)$$

where N_W is population size in the watersheds of some county, C, and N_C is the total new population in that county, estimated as the difference between number of people in the county in 2025 and 1995. A_C is the total developable area (ha) in each county and A_W is the total watershed area (ha) in each county. This is a conservative assumption, since demographic statistics suggest a preference for development in proximity to water. Thus, we underestimate the pressure on the watershed.

The amount of watershed area needed for development depends on urban typology, notably the amount of land that is urbanized per person and the number of people that will reside in the watershed, N_W. The urban land use ratio (ULUR), i.e., the urban land use per person in a particular place (e.g., town, county), expresses the former attribute and varies with urban typology. Suburban typologies are land intensive relative to traditional typologies. A ULUR of 3, meaning that three acres (or 1.22 hectares; the ULUR is given in acres, but all computations were performed using the metric system) of urban land are required to support each person, was used to estimate the suburban land requirement, A_S, in the watersheds. This ratio is conservative relative to ULUR values (5–8) in some rapidly developing counties in coastal South Carolina (BCD-COG 1996; Allen and Lu 2003). The watershed area required by the traditional urban typology, A_T, was computed using a ULUR of 1 (= 0.42 hectares/person). This is also conservative, as ULURs in traditional small towns and cities, such as Charleston, typically range from 0.05 to 0.1 (Kleppel et al. this volume). The proportion of the total watershed area in each county, A_W, required for development is A_X/A_W, where urban typology, X, is either suburban, S, or traditional, T.

The benthic macroinvertebrate species density in a tidal creek in a watershed of average area, A_M, in county C, resultant from suburban or traditional urban development was estimated from the SAR [Equation (4.1)],

$$S'_C = [(A_N/A_W) (c_N A_M^{ZN})] + [(A_X/A_W) (c_X A_M^{Zx})] \qquad (4.3a)$$

$$= S'_N + S'_X, \qquad (4.3b)$$

where S'_C, S'_N, and S'_X are, respectively, the average species density for tidal creeks in county, C, in undeveloped (N) and urbanized (X) portions of the average watershed of area, A_M (= 126.3 ha). X can be either the sub-urban (S) or the traditional urban (T) typology. The scaling factor, c, and slope, z, are specific to undeveloped (N) and urbanized (X) watersheds. The ratio A_N/A_W refers to the proportion of the total watershed area in the county, A_W that is undeveloped. A_X/A_W is as in Equation (4.2), above.

The proportional change in S'_C (i.e., $\Delta S'_C$) between 1995 and 2025 was computed with the equation

$$\Delta S'_C = (S'_{C2025} - S'_{C1995}/S_N), \qquad (4.4)$$

where $S'_{C1995} = S'_N$. That is, in 1995, the watersheds in this analysis were not yet urbanized. $S'_{C2025} = S'_C$ in Equation (4.3(a)).

4.3 Results

4.3.1 Species-Area Relationships and Watershed Urbanization

The relationship between tidal creek benthic macroinvertebrate species density, S', and watershed area, A, depends on land use (Figure 4.2). While the SARs for both undeveloped and urbanized watersheds conform to a power law and both slopes (z) approach 0.25 (see MacArthur and Wilson 1967; but also, Crawley and Harral 2001), the constant c is approximately 1.8 times higher in undeveloped watersheds than in urbanized watersheds. Thus, at realistic values of A (50–500 ha), S' in undeveloped watersheds is approximately double that in undeveloped watersheds.

4.3.2 Demographic Trends and Changing Land Cover in Coastal Counties

Between 1995 and 2025 projected in-migration to the coastal counties of South Carolina will exceed 520,000 people. We estimate that more than 450,000 people will live in small coastal watersheds associated with

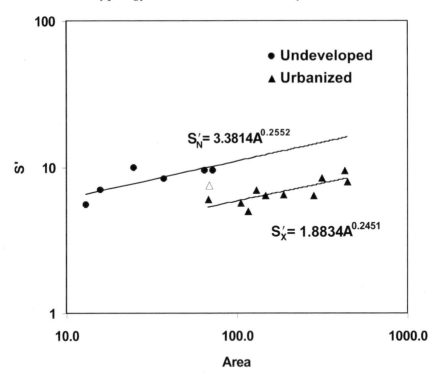

Fig. 4.2. Species-area relationships for benthic macroinvertebrate species density, S', versus watershed area, A, in hectares. S' was assessed in tidal creeks in undeveloped and urbanized watersheds near Charleston, South Carolina, as part of the Tidal Creeks Study (Holland et al. 1996). Circles represent undeveloped watersheds (species density = S'_N), n = 6. Triangles represent watersheds dominated by urban development (species density = S'_X), n = 11. Open triangle is an anomalous point and was omitted from the analysis. Data sets were fit by a power law per Equation (4.1)

the salt marsh–tidal creek complex. Population growth rates will vary substantially among counties. Horry, Berkeley, Dorchester, and Beaufort counties will experience increases of more than 75 percent, while Georgetown, Charleston, Colleton, and Jasper counties will grow by less than 50 percent.

4.3.3 Urban Development in Coastal Watersheds and Benthic Biodiversity in Tidal Creeks

In Table 4.3(A-G), we step through the calculations used to estimate average benthic species density in coastal tidal creeks in 2025. Based

Table 4.3. Calculation of Predicted Changes in Tidal Creek Benthic Macroinvertebrate Species Density in Eight Coastal Counties of South Carolina as a Function of Population Growth in Adjacent Watersheds Between 1995 and 2025.

A. Population in Thousands

Year	Georgetown	Horry	Charleston	Berkeley	Dorchester	Colleton	Beaufort	Jasper
1995	51	158	288	135	85	37	95	17
2025	70	309	370	262	127	46	183	19
Growth (N25–N95)	19	151	82	127	42	9	88	2

B. New population of county watersheds in 2025 [$N_w = N_c(A_w/A_c)$]

Georgetown	Horry	Charleston	Berkeley	Dorchester	Colleton	Beaufort	Jasper
13435	122949	66838	66502	6595	3350	174505	671

C. Watershed Area Required ($A_x = ULUR \times N_w$)

	Georgetown	Horry	Charleston	Berkeley	Dorchester	Colleton	Beaufort	Jasper
Suburban	16390	149997	81542	81133	8046	4088	212897	819
Traditional	5441	49794	27069	26933	2671	1357	70675	272

(*Continued*)

Table 4.3. Calculation of Predicted Changes in Tidal Creek Benthic Macroinvertebrate Species Density in Eight Coastal Counties of South Carolina as a Function of Population Growth in Adjacent Watersheds Between 1995 and 2025. (Continued)

	Georgetown	Horry	Charleston	Berkeley	Dorchester	Colleton	Beaufort	Jasper
D. Proportion of Total Watershed Area Needed (A_X/A_W)								
Suburban	0.13	1.00	0.42	1.00	0.40	0.05	1.00	0.02
Traditional	0.04	0.44	0.14	0.34	0.13	0.02	0.45	0.01
E. S' for Developed Portions of Watersheds [$S'_X = (A_X/A_W)c_X A_M^{Zx}$]								
Suburban	0.77	6.17	2.60	6.17	2.48	0.32	6.17	0.11
Traditional	0.26	2.71	0.86	2.12	0.82	0.11	2.77	0.04
F. S' for Undeveloped Portions of Watersheds [$S'_N = (1-A_X/A_W)c_N A_M^{ZN}$]								
Suburban	10.18	0.00	6.74	0.00	6.95	11.02	0.00	11.42
Traditional	11.15	6.53	10.00	7.64	10.08	11.43	6.41	11.56
G. Average S' for County C ($S'_C = S'_N + S'_X$)								
Suburban	10.95	6.17	9.33	6.17	9.43	11.35	6.17	11.53
Traditional	11.40	9.23	10.87	9.75	10.90	11.53	9.18	11.60

on projected population growth (Table 4.3(A)) and the distribution of a portion of that population within the small coastal drainages that we define as watersheds (Table 4.3(B)), an average of 20–60 percent of the watershed area in the coastal counties will be required to support the incoming population, depending upon typology, i.e., ULUR (Table 4.3(C, D)). Suburban development (ULUR = 3) along the coast will, by 2025, require the urbanization of roughly one-half million ha of land in the watersheds to accommodate the expected coastal population growth. In Horry, Berkeley, and Beaufort Counties the suburban typology would require more than 100 percent of the watershed area in each county. [When the estimate of A_x/A_w exceeded 100 percent, a default value of 1.00 was input (Table 4.3(D)).] Thus, with a suburban growth scenario, the watershed areas of Horry, Berkeley, and Beaufort Counties are too small to accommodate the entire in-migrant population expected by 2025 (Figure 4.3(a)). Conversely, traditional urban development, with a ULUR = 1, would use about one-third of the space required by the suburban typology.

4.3.4 Typological Influences on Benthic Biodiversity

The suburban scenario predicts that on average, benthic macroinvertebrate species density in coastal South Carolina's tidal creeks will, by 2025 (Table 4.3(A-G)) be reduced to 76 percent (i.e., a loss of 24 percent), of what it was in 1995 (Figure 4.3(b)). In the three fastest-growing counties (Horry, Berkeley, Beaufort), S'_C will decline by 38 percent. A substantially smaller impact, an all-county average reduction in S'_C of 9 percent and a reduction in the three fastest-growing counties of 16 percent, is predicted if, instead of suburban typologies, traditional urban typologies are used to develop in coastal watersheds.

4.4 Discussion

4.4.1 Benthic Tidal Creek Species-Area Relationships in Typologically Different Watersheds

It is well known that urbanization in watersheds affects biotic community structure in the aquatic systems that drain them (Valiella and Teal 1979; Limburg and Schmidt 1990; Scheuler 1995). In the present study, it would appear that the species-area relationship was sensitive

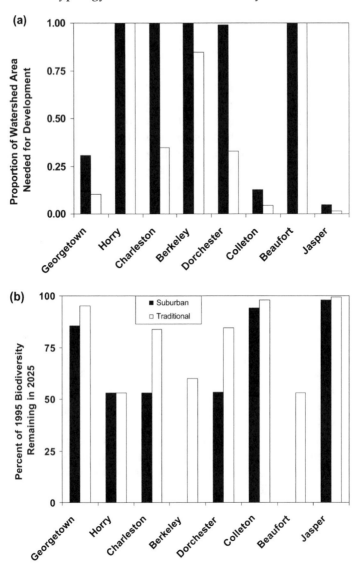

Fig. 4.3. (a) Proportion of watersheds in South Carolina's coastal counties that will be developed between 1995 and 2025 as a function of different urban typologies. For the suburban typology (black bars), the urban land use ratio (ULUR) is 3. For the traditional urban typology (white bars) the ULUR is 1. (b). Percent of the benthic macroinvertebrate biodiversity, species density, S', in tidal creeks in South Carolina's coastal counties in 2025 relative to that in 1995

to the influence of watershed urbanization on benthic invertebrate communities. Tidal creeks in urbanized watersheds support about half the biodiversity of creeks in undeveloped watersheds.

Clearly, SARs do not reflect the fine details of spatiobiotic relationships, and our analysis does not permit detailed examination of the influence of species variability at seasonal or other biologically relevant time scales. However, despite their coarseness, relationships identified here can help elucidate general trends and create a basis for comparing alternative development styles in the coastal zone.

4.4.2 Coastal Growth—Implications for Tidal Creek Habitat and Benthic Biodiversity

We have emphasized throughout this paper that our projections are intended to reflect predicted trends rather than precise outcomes. However, the trends we predict are, when evaluated empirically, supported by observations. Lerberg (1997) described reductions in tidal creek benthic macroinvertebrate species richness accompanying intensive watershed development, and Lerberg et al. (2000) noted that tidal creeks in urbanized watersheds were dominated by a few species of oligochaetes. Similar observations have been made in coastal and estuarine ecosystems that drain other urbanized landscapes (Weinstein 1996). Numerous studies suggest that ecosystem functionality is linked to biodiversity (Hooper and Vitousek 1997; Tilman et al. 1997; Kinzig et al. 2001). Reductions in benthic biodiversity may reduce functionality (Pimm 1984; Kondoh 2003) and increase susceptibility to biological invasion (Elton 1958; Mack et al. 2000; but see Simberloff 1995).

Two lessons emerge from this work. First, development along the coast, if allowed to continue according to current suburban design standards, will result in the urbanization of a significant proportion of the watersheds in South Carolina's coastal counties, accompanied by an equally significant loss of the benthic biodiversity in the tidal creeks of these watersheds. The second lesson is that typology can alter these outcomes. By employing designs like those of traditional small towns, cities and increasingly popular new urbanist communities [which, data suggest, often perform well in the market (Eppli and Tu 1999)], projected in-migrant population growth can be supported on a third of the land required by suburbs, and the decline in S'_c can be reduced considerably.

Acknowledgments

We are grateful to L. Pomeroy, G. Robinson, A. F. Holland, and S. Lerberg for helpful discussions and comments on the various drafts of the manuscript. Support for this project was provided by the Land Use-Coastal Ecosystem Study through NOAA grant NA960PO113 from the South Carolina Sea Grant Consortium and the NSF Biocomplexity Research Program under grant DEB0083325.

References

Allen, J. and K. S. Lu. 2003. Modeling and prediction of future urban growth in the Charleston region of South Carolina: A GIS-based integrated approach. *Conservation Ecology* 18: http://www.consecol.org/18/iss2/art2.

Arendt, R. 1992. Rural by Design. American Planning Association, Chicago, IL.

Arrenhius, O. 1921. Species and area. *Journal of Ecology* 9:95–99.

BCD-COG (Berkeley, Charleston, Dorchester Council of Governments). 1996. Twenty Year Time Series Analysis of Satellite Remote Sensor Data to Detect Environmental and Developmental Change along Coastal Georgia and South Carolina. South Carolina Department of Natural Resources, Columbia. (Available in CD-format.)

Cai, W.-J. and L. Pomeroy. 1999. State of knowledge of respiratory processes and net productivity of intertidal marshes of Georgia and South Carolina. Coastal Ocean Program, U.S. Department of Commerce, NOAA Coastal Ocean Program, Land Use-Coastal Ecosystem Study, Charleston, SC. 23p. http://www.luces.org/documents/SOKreports/respiratoryproc_netproduct.pdf.

Collins, J. P., A. Kinzig, N. B. Grimm, W. F. Fagan, D. Hope, J. Wu and E. T. Borer. 2000. A new urban ecology. *American Scientist* 88:416–425.

Crawley, M. J. and J. E. Harral. 2001. Scale dependence in plant biodiversity. *Science* 291:864-868.

Culliton, T. J., M. A. Warren, T. R. Goodspeed, D. G. Remer, C. M. Blackwell, and J. J. I. Mc Donough. 1990. Fifty years of Population Change along the Nation's Coasts 1960–2010. U.S. Department of Commerce, NOAA. Rockville, MD. 41p.

Duany, A., E. Plater-Zyberk, and J. Speck. 2000. Suburban Nation. North Point Press, New York, NY. 293p.

Eisenberg, E. 1998. The Ecology of Eden. Vintage, Random House, New York, NY. 612p.

Elton, C. S. 1958. The Ecology of Invasions by Animals and Plants. Methuen, London. 181p.

Eppli, M. J. and C. C. Tu. 1999. Valuing the New Urbanism. Urban Land Institute, Washington, DC. 86p.

Holland A. F and 12 others. 1996. The Tidal Creek Project. Interim report. Charleston Harbor Project. South Carolina Department of Natural Resources, Charleston, SC. 229p.

Hooper, D. U. and P. M. Vitousek. 1997. The effects of plant composition and diversity on ecosystem processes. *Science* 277: 1302–1305.

Hoss, D. E. and D. W. Engel. 1996. Sustainable development of the southeastern coastal zone: Environmental impacts of fisheries, pp. 221–240. In, Vernberg, F. J., W. B. Vernberg, and T. Siewicki (eds.), Sustainable Development in the Southeastern Coastal Zone. University of South Carolina Press, Columbia.

Katz, P. 1994. The New Urbanism. McGraw-Hill, New York, NY. 245p.

Kinzig, A. P., S. W. Pacala, and D. Tilman (eds.). 2001. The Functional Consequences of Biodiversity. Princeton University Press, Princeton, NJ. 365p.

Kleppel, G. S. 2002. Urbanization and environmental quality: Implications of alternative development scenarios. *Albany Law Environmental Outlook Journal* 8: 37–64.

Kleppel, G. S., R. Becker, J. S. Allen, and K.-S. Lu. (This volume.) Trends in land use policy and development in the coastal Southeast. In, Kleppel. G. S., M. R. DeVoe, and M. V. Rawson (eds.), Implications of Changing Land Use Patterns to Coastal Ecosystems. Springer-Verlag, New York, NY.

Kleppel, G. S. and M. R. DeVoe. 2000. The people factor. In, Davis, J. E. (ed.), A Sense of Place. *South Carolina Wildlife* 47:32–43.

Kleppel, G. S., S. Madewell, and S. E. Hazzard. 2004. Responses of emergent marsh wetlands in upstate New York to variations in urban typology. *Ecology and Society* 9:1. 18p. http://www.ecologyandsociety.org/vol9/iss5/art1.

Kondoh, M. 2003. Foraging adaptation and the relationship between food-web complexity and stability. *Science* 299: 1388–1399.

Lerberg, S. B. 1997. Effects of watershed development on macrobenthic communities in the tidal creeks of the Charleston Harbor estuary. Master's thesis, University of Charleston, Charleston, SC. 257p.

Lerberg, S. B., A. F. Holland, and D. M. Sanger. 2000. Responses of tidal creek macrobenthic communities to the effects watershed development. *Estuaries* 23: 838–853.

Limburg, K. E. and R. E. Schmidt. 1990. Patterns of fish spawning in the Hudson River watershed: Biological response to an urban gradient? *Ecology* 71:1238–1245.

Lowrance, R. R. 1998. Riparian forest ecosystems as filters for nonpoint-source pollution, pp. 113–141. In, Pace, M. L. and P. M. Groffman (eds.), Successes, Limitations and Frontiers in Ecosystem Science. Springer-Verlag, New York, NY.

MacArthur, R. H. and E. O. Wilson. 1967. The Theory of Island Biogeography. Princeton University Press, Princeton, NJ. 203p.

Mack, R. N., D. Simberloff, W. M. Lonsdale, H. Evans, M. Clout, and F. A. Bazzaz. 2000. Biotic invasions: Causes, epidemiology, global consequences and control. *Ecological Applications* 10:689–710.

Nixon, S. 1980. Between coastal marshes and coastal waters—a review of twenty years of speculation and research on the role of salt marshes in estuarine productivity and water chemistry, pp. 437–525. In, Hamilton, P. and K. B. Macdonald (eds.), Estuarine and Wetland Processes with Emphasis on Modeling, Vol. 11. Plenum, New York, NY.

Nixon, S. W. 1995. Coastal eutrophication: A definition, social causes and future concerns. *Ophelia* 41:237–249.

Paerl, H. W. 1988. Nuisance algal blooms in coastal, estuarine and inland waters. *Limnology and Oceanography* 33:823–847.

Paerl, H. W. 1997. Coastal eutrophication and harmful algal blooms: Importance of atmospheric deposition and groundwater as sources of "new" nitrogen and other nutrients. *Limnology and Oceanography* 42:1154–1165.

Pennings, S. C. and M. D. Bertness. 1999. Using latitudinal variation to examine effects of climate on coastal salt marsh pattern and process. *Current Topics in Wetland Biogeochemistry* 3:100–111.

Pimm, S. L. 1984. Complexity and stability of ecosystems. *Nature* 307:321–326.

Pimm, S. L. and P. Raven. 2000. Extinction by the numbers. *Nature* 843–845.

Plotkin, J. B. et al. 2000. Predicting species diversity in tropical forests. *Proceedings of the National Academy of Sciences* 97:10850–10854.

Porter, D. E., W. K. Michener, T. Siewicki, D. Edwards, and C. Corbett. 1996. Geographic information processing assessment of the impacts of urbanization on localized coastal estuaries: A multidisciplinary approach, pp. 355–388. In, Vernberg, F. J., W. B. Vernberg, and T. Siewicki, (eds.), Sustainable Development in the Southeastern Coastal Zone. University of South Carolina Press, Columbia.

Preston, F. W. 1962. The canonical distribution of commonness and rarity. Part I. *Ecology* 43:185–215; Part II. *Ecology* 43:410–432.

Sala, O. E. et al. 2000. Global biodiversity scenarios for the year 2100. *Science* 287: 1770–1774.

Scheuler, T. 1995. The importance of imperviousness. *Watershed Protection Techniques* 1: 37-48.

Simberloff, D. 1995. Why do introduced species appear to devastate islands more than mainland areas? *Pacific Science* 49:87–97.

Tilman, D. et al. 1997. The influence of functional diversity and composition on ecosystem processes. *Science* 277: 1300–1302.

U.S. Census Bureau. 2000. South Carolina census data. Department of Commerce, Bureau of the Census, http://www.census.gov.

Valiela, I. and J. M. Teal. 1979. The nitrogen budget of a salt marsh ecosystem. *Nature* 280: 652–656.

Vernberg, F. J., W. B. Vernberg, and T. Siewicki (eds.). 1996. Sustainable Development in the Southeastern Coastal Zone. University of South Carolina Press, Columbia. 519p.

Vitousek, P. M. 1994. Beyond global warming: Ecology and global change. *Ecology* 75:1861-1876.

Weinstein, J. F. 1996. Anthropogenic impacts on salt marshes—a review, pp. 135–170. In, Vernberg, F. J., W. B. Vernberg, and T. Siewicki, (eds.), Sustainable Development in the Southeastern Coastal Zone. University of South Carolina Press, Columbia.

Wenger, S. 1999. A review of the scientific literature on riparian buffer width, extent and vegetation. Institute of Ecology, University of Georgia, Athens. 57p.

Part II

Coastal Hydrology and Geochemistry

The Relationship of Hydrodynamics to Morphology in Tidal Creek and Salt Marsh Systems of South Carolina and Georgia

Jackson O. Blanton, Francisco Andrade, and M. Adelaide Ferreira

Summary by Jackson O. Blanton and Daniel R. Hitchcock

Salt marshes, typical of those found in South Carolina and Georgia, are covered and uncovered approximately twice daily by the rise and fall of the tide. Salt water is transported to the marshes by an array of tidal creeks ranging from 0 to 15 meters in depth and 5 to 500 meters in width. The water fills the creeks on the rising tide and then spills out into the marsh. As the tides fall, the creeks, starting with the smallest, collect the water much as do the capillary blood vessels in mammals, and convey it to ever-larger creeks by which it is eventually transported into the sounds and estuaries. The banks of tidal creeks undergo considerable sediment erosion, while marsh surfaces are areas of sediment deposition, allowing them to keep pace with sea level rise.

When the marshes and creeks are treated as a whole, one sees that the water surface area changes from its smallest value at spring tide low water to its maximum at spring tide high water. The areas encompassed by neap tide fit somewhere in between. Thus, one can derive curves to estimate the variation of the water surface area in the salt marsh-tidal creek complex as a function of water depth. These curves, called hypsometric curves, are needed to estimate the flushing and flooding potential of a marsh.

To understand how the hypsometry of a salt marsh system can be determined, one can use an example from a small system of tidal creeks in and around the Okatee River in South Carolina. A series of six airborne aerial images was used to measure flooded water area during the flood phase of the tide. The water level was measured with a pressure sensor at the bottom of the main creek, thus providing a vertical reference level for the corresponding flooded area. The shape of this curve is governed by the morphology of the creek-marsh areas, i.e., distribution of wetted surface area as water level increases in the area of interest.

The change in area as depth increases represents a certain volume that covers the marsh from its low-water stage to its high-water stage. This volume, defined as the tidal prism, equals the volume of water that flows through a cross-section of the tidal creek. The growth rate of the prism is equal to the velocity flowing through the cross-section of the creek (an equality based on a principle of fluid dynamics called continuity). The faster the horizontal area increases, the larger the transport of water must be to support the growth of the surface area. While the tide is largely responsible for increases in the area (and depth) covered, water level changes can also be induced by winds along the coast. Winds that blow coastal water to the south pile up water in the estuaries and sounds, and water level there can increase by 0.5 m or so. Wind from the opposite direction can depress water levels by a comparable amount.

There are several management issues associated with strong tidal currents flowing through the creeks. For example, the creation of docks and marinas along a tidal creek will impede flow where impediments did not previously exist. Boats comfortably afloat at the beginning of the project wind up becoming stuck in mud during certain parts of the tidal cycle several years later. The most common solution is to implement a dredging program to remove sediment that was being removed by the high tidal flow that existed before development. Thus, the economic service provided by nature must be subsidized through an expensive dredging program in order to keep the waterway usable. Alternatively, by developing a hypsometric profile for its waterways before development, a community can guide the siting of projects such as marinas, thereby reducing costs to taxpayers in the future.

Using principles outlined by Friedrichs and Perry (2001) and Pethick (2001), it seems prudent that planning for development in and adjacent to tidal marshes take account of the natural balance between the tidal transport of water and sediment supply. This means that (1) the heavy marsh grass density must be maintained in order to provide an environment favorable to the settling of suspended sediments, (2) that the natural sediment supply from land should be maintained, and (3) the inland migration of the marsh as sea level rises should be anticipated and planned for appropriately.

5.1 Introduction

Salt marshes in Georgia and South Carolina are alternately flooded and drained by the semidiurnal tide. The marshes are usually covered with *Spartina* grasses but also have some mudflats. Some areas have extensive freshwater tidal marshes. While many tidal creeks have freshwater sources, many drain the interior salt marshes, which have limited fresh water sources. Many areas of the marsh are poorly flushed in areas with dense marsh vegetation.

Tidal creeks are typically areas of high sediment erosion, while the tidal marshes themselves usually have high deposition (Friedrichs and Perry 2001). High deposition rates allow the marsh to react to sea level rise by increasing its elevation and migrating landward (Pethick 2001). Tidal creeks serve as important conduits that transport material into and out of the marshes. Many tidal creeks connect to major estuaries. Others may connect two different estuarine regimes, and a tidal creek node in the connecting creek may limit flow between the two regimes (Schwing and Kjerfve 1980). Tidal marshes and creeks are particularly responsive to changes in freshwater supply from agents such as groundwater, adjacent estuaries, and local rainfall. Yet we know little about the sensitivity of this response.

This paper begins with a description of the morphology of salt marshes in South Carolina and Georgia. The paper focuses on the geological setting and the morphological properties that link directly to hydrodynamics and tidal circulation. After a description of the tidal regime typical of coastal areas in South Carolina and Georgia, the Okatee River in South Carolina will serve as a "prototype" tidal creek to demonstrate the link between tidal circulation and creek-marsh morphology. The effect that wind stress on the continental shelf has on water level and exchange with the ocean will be described. Finally, the relevance of tidal hydrodynamics and marsh sedimentation to management issues will be discussed.

5.2 Links Between Hydrodynamics and Morphology

Hydrology of the salt marsh, the area and duration of flooding, frequency of inundation, and permeability of sediments determine nutrient and particulate transport within the marsh. One study of interstitial water residence time yielded values of 1–26 hr for the upper 10 cm of sediments (Bollinger and Moore 1993). Throughout the estuary, reactive nutrients in the marsh, such as ammonium and phosphate, exhibit quite different salinity-nutrient mixing diagrams during spring tides, when marshes flood, versus neap tides that do not inundate the marsh surface (Vörösmarty and Loder 1994). The flooded area and marsh depth below spring high tides can thereby greatly influence the whole estuarine ecosystem.

Salt marshes keep pace with rising sea level by accumulating sediment at rates of 1–10 mm per year (Wolaver et al. 1988) and migrating landward at a rate as high as 1m/year for a surface slope of 1:1000 (Pethick 2001). Sediments can arrive at the marsh surfaces through resuspension and transport by strong tides and storms, as well as by river flow. Time scales associated with resuspension are associated with the fortnightly tides and similar time scales associated with storms. Nearby rivers can also supply large amounts of suspended sediments during flood stages, with time scales that are usually much longer. Regardless of source, the mechanism of sediment delivery is not obvious. The simple model is that sediment is delivered and deposited by high tidal flooding. When marshes are flooded by waters carrying a large sediment load, the shallow depths and obstructions caused by grass stalks cause the flow to slow and allow sediments to settle (Leonard and Reed 2002). Much of this sediment can be subsequently mobilized and removed in a single rain event occurring at low tide (Chalmers et al. 1985). However, in some cases much of the annual sediment accumulation may occur during one or two storm events (Stumpf 1983). Tidal range and fortnightly variations would be expected to have a great influence on the effects of single events on sediment accumulation and erosion in the salt marsh—tidal creek complex.

Estuarine morphology bears a strong relationship to tidal circulation in estuaries and tidal creeks (Postma 1967; Healey et al. 1981; Zarillo 1985; Dronkers 1986). Ebb-dominant estuaries are common south of Cape Romain, South Carolina (Friedrichs and Aubrey 1988), due to the inefficient water exchange near the time of high water between the extensive intertidal marsh areas and the deep channels. Friction slows the propagation of high tide through shallow grassy intertidal creeks and marshes, which prolongs the duration of flood tide. Similarly, friction retards the fall of the tide relative to the fall in the open channels, thus providing a strong surface slope that drives a faster ebb tide but for a shorter duration (Dronkers 1986). Numerical studies confirm that channels without intertidal storage in flats and marshes are flood dominant, while channels with large areas of intertidal storage are ebb dominant (Speer and Aubrey 1985). Zarillo (1985) clearly demonstrated that in the ebb-dominated Duplin River (Georgia), threshold values for sand transport were exceeded for two hours or more during ebb tide, resulting in seaward migration of coarse-grained bed forms consistent with ebb dominance. The changes in marsh morphology along the various levels of tidal creeks (levels that become shallower and are fed by smaller creeks) change the character of the tidal currents due to continuity and bottom and side friction (Healey et al. 1981) in a manner that alters the size fraction and direction of sediment transport (Dronkers 1986).

The geometry of tidal creeks is hydraulically controlled by depth and flow. This control is governed by strong feedback involving resuspension, erosion, deposition, and river supply. Sediment resuspension during strong spring tides creates erosive bed conditions, but only fine sediments are resuspened during neap tides (Gardner and Bohn 1980; Smith and McLean 1984). Erosion plus resuspension on the beds of the creeks are important to the dispersion of benthic meiofauna, including some of the larval stages of fish and shellfish (Palmer and Güst 1985). Thus the fortnightly cycle is very important to the dynamics of tidal creeks. In many systems large erosion may occur with storm surges, but because the duration of such events is short, they are thought to have little total effect in Georgia and South Carolina marshes over seasonal time scales (Gardner et al. 1992).

5.3 Tidal Circulation and Morphology: Example from the Okatee River, South Carolina

The link between the tidal regime and morphology is well accepted by most. To establish this link for an area undergoing rapid development, a series of six airborne aerial images of a small tidal creek in South Carolina (Figure 5.1) was used to measure flooded water area during the flood phase of the tide (Andrade et al. in preparation). The water level, measured with a sub-surface pressure gage in the study area, was used as the vertical reference level for the corresponding flooded area. A hypsometric curve was defined from this data set that yielded water surface area as a function of water level.

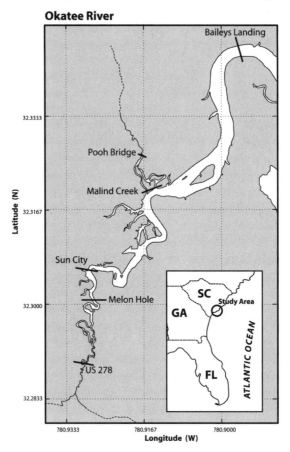

Fig. 5.1. Location of Okatee Creek (inset) and detailed map of the study area. The lines mark the locations upstream of which the hypsometric curves were calculated. The subsurface pressure gauge used to monitor water level during the overflights was located at Melon Hole

5.4 Examples of the Hypsometric Curve for the Okatee River System

The hypsometric curve for the entire study area (Figure 5.2) shows that water surface area increased relatively slowly as the water level increased from low water (LW) to one-third of the way to high water (HW). Thereafter, water surface area increased at a faster rate. The

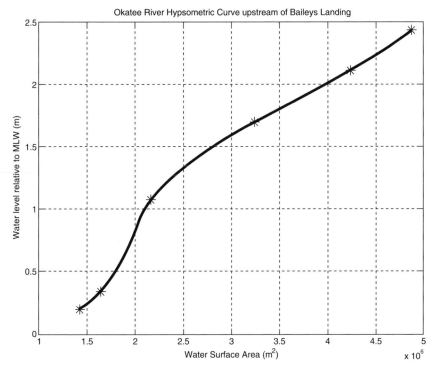

Fig. 5.2. The hysometric curve for the Okatee River upstream of Baileys Landing. The six values obtained from the aerial images are denoted by asterisks and the curve results from a spline fit to these values. Note that water surface area increased by more than a factor of three over the flood phase of the tide. See Figure 5.1 for location

growth of water area over time influences the volume transport of the tidal current during flood through a down-stream cross-section of a given area.

We compare two small upstream intertidal areas in Okatee River and Malind Creek (Figure 5.3). The curve for upper Okatee River shows the same general properties of the other sections of Okatee River. However, the upper part of Malind Creek (a tidal creek that feeds the Okatee River) looks quite different. The hypsometry suggests steep banks with small water surface areas until mid-tide after which the area increases at a fast rate reflecting the very low slope of the high marsh areas.

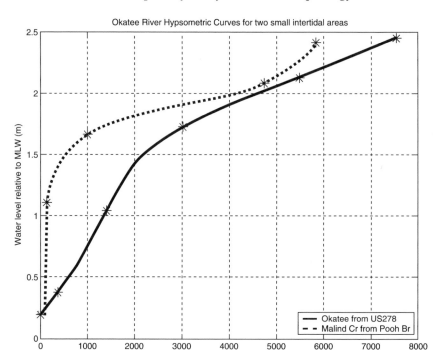

Fig. 5.3. Hypsometric curves for the intertidal areas upstream of U.S. 278 and Pooh Bridge (Malind Creek). See Figure 5.1 for location. The six values obtained from the aerial images are denoted by symbols and the curves result from a spline fit to these values. Note that the scale of the x-axis is smaller than that of Figure 5.2 by almost three orders of magnitude

5.5 Tidal Currents Derived from Hypsometric Curves

The tidal current was calculated from the hypsometric curve using the continuity equation to conserve volume. The average current through a cross-section is related to the cross-sectional area and the tidal prism upstream of the cross-section through the equation of continuity:

$$u A_z = A_{xy} \frac{\partial \eta}{\partial t} \tag{5.1}$$

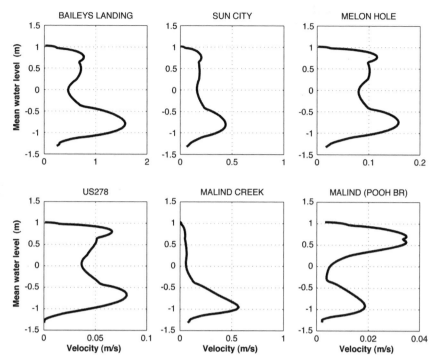

Fig. 5.4. Currents derived from the hypsometric curves at locations shown in Figure 5.1

where u is the cross-sectionally averaged axial velocity, A_z is the cross-sectional area at the downstream end where the hypsometric curve applies, A_{xy} is the horizontal surface area of the estuary's water surface, and $\partial\eta/\partial t$ is the rate of change of water level. Both A_z and A_{xy} are functions of time and were calculated for all cross-sections based on the digital elevation model derived from the infrared images of the Okatee River system (Andrade et al. in preparation).

Applying Equation 1 to six sections of the Okatee River system (Figure 5.1) yielded a series of flood current speed as a function of water level (Figure 5.4). The fastest flood currents generally occurred early in the flood phase when rate of change of water area upstream of the measurement was maximum. Also, note that there are two maxima in the flood current at most areas, one soon after water level began to rise, the other closer to high water.

The calculated currents were compared with observations at two of the smallest tidal water sheds (Figure 5.5). While these records were

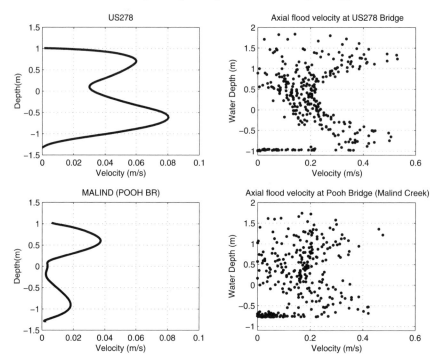

Fig. 5.5. Comparison of currents calculated from Equation (5.1) with field measurements. (a) Calculations for U.S. 278; (b) Measurements at U.S. 278; (c) Calculations for Pooh Bridge; (d) Measurements at Pooh Bridge

not obtained at the same time as the aerial images, many features of the field data can be discerned in the currents derived from Equation (5.1). The field measurements confirm the two peaks in flood velocity calculated from the hypsometric curve, but the difference in magnitude of the two peaks are not verified in the field data. As expected, the field measurements of velocity obtained in the deepest part of the channels were larger than the cross-sectional mean.

5.6 Effect of Low-Frequency Phenomena from Continental Shelves on Tidal Creeks

Water-level fluctuations with periods greater than 2–3 days (subtidal frequencies) can be driven by wind-stress (Ekman fluxes) on the continental shelf (Klinck et al. 1981) or by remotely forced shelf waves

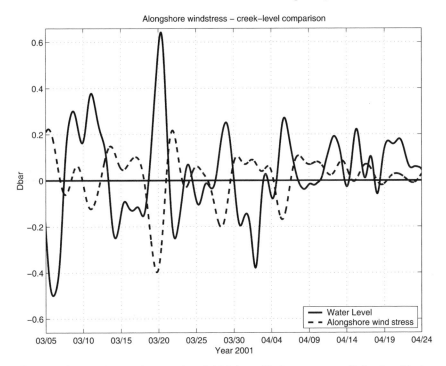

Fig. 5.6. Comparison of water level (tidal oscillations removed) in the Okatee River with wind stress measured at a buoy on the continental shelf off Charleston, South Carolina (NDBO Buoy No. 41006)

(Schwing et al. 1988). These fluctuations are superimposed on the normal tidal fluctuations and cause flooding and drying of the marshes over longer time scales. Estuarine exchange with the ocean is maximized under these conditions. On the other hand, cross-shelf stress can transport and redistribute water inside the estuary while causing relatively little exchange with the ocean.

We demonstrate the magnitude of such sub-tidal fluctuations in water level by comparing water level fluctuations inside Okatee River with along-shore wind stress on the adjacent continental shelf (Figure 5.6). Note the visual correspondence of along-shelf wind stress with water level. The downwelling-favorable stress after 16 March caused water level to increase to 0.8m. Similar water level increases were observed for the two downwelling events of 28 March and 7 April. These caused water level increases of 0.3–0.4 m. Upwelling wind stress caused the opposite effect, but water level excursions were less for a unit of wind stress. The important point here is that when downwelling wind conditions that increase sea level at the coast coincide with spring tide, sea level can increase by as much as 1 m, thereby inundating even larger intertidal areas.

5.7 Some Management Issues

The large SM-TCC systems along the South Carolina-Georgia coast have a mesotidal environment along the tidal creeks. Here the tidal currents are the principal agent governing the deposition and resuspension of inorganic sediments (Friedrichs and Perry 2001). By contrast, areas of the marsh adjacent to land may have smaller tidal ranges because of the build-up of surface elevation. It is in these areas that storms and floods (much more erratic and unpredictable than the tides) become the important agents maintaining sediment supply.

We have used the Okatee River in South Carolina to illustrate the physical oceanographic attributes of mesotidal salt marshes. The Okatee shares features that are typical of many salt marshes throughout the world. One over-arching principle that connects these mesotidal marshes is that marshes that are in their "natural" state are approaching, if not at, "a dynamic equilibrium with sediment supply, vegetative growth, and relative sea level, rather than far out of equilibrium on a slow evolution toward geological maturity" (Friedrichs and Perry 2001). Sediment supply is controlled by source concentration, distance from the source, and duration of inundation which is related to the hypsometric curve. Deposition rates are higher for longer inundation periods, and the higher rates are expected to occur during periods of accelerated sea level rise. The additive effect of *proximity to sediment source* and *duration of inundation* leads to a relatively uniform accretion of inorganic sediments along the berms at the boundary between a tidal creek and the surrounding marsh. Moreover, as pointed out by Pethick (2001), the vertical rise in marsh surface results in a *horizontal* migration of the marsh landward. The low slope characteristic of marshes in the southeastern U.S. would be particularly sensitive to this migration.

Marshes that are undergoing changes associated with alterations of land use may rapidly evolve away from dynamic equilibrium (Friedrichs and Perry 2001). This implies that development in and adjacent to salt marshes should take account of the dynamic equilibrium issues outlined in the previous paragraph. This requires that land-use designs recognize and incorporate both the vertical and horizontal components of marsh migration (Pethick 2001). Moreover, maintaining the natural high density of marsh grasses assures a low-velocity environment that maintains a high rate of sedimentation. Any development that reduces the sediment supply to the marshes (like hardening the landscape adjacent to marshes, constructing dams, and creating water diversion projects) and does not allow room for the marsh to shift landward will be subject to high maintenance costs.

The maintenance costs associated with land use changes need further comment. One may argue that environments that are in "dynamic equilibrium" maintain the optimum balance between import and export of material. A land use project should, as one of its fundamental objectives, disturb this balance in as small a way as possible—not for aesthetic reasons, although these are important—but for economic reasons. For example, the creation of docks and marinas along a tidal creek will create a high-friction environment where none existed before: the flow of water will be slowed, sedimentation will increase, boats that were comfortably afloat at low water become stuck in the mud at low water. The most common solution is to implement a dredging program to remove sediment that was once being removed by the high tidal flow present before development.

The tidal flow that used to maintain relatively deep water at the creek's edge performed a economic service as a component of the dynamic equilibrium of the system. This service must now be performed by an expensive dredging program in order to keep the waterway usable. Who now bears the cost of the dredging program? In most cases, the developer is no longer on the scene. The cost then falls either to local governments or to the landowners themselves. Good land-use planning should examine carefully who will ultimately bare the maintenance costs long after the development has been completed.

5.8 Conclusions

We have shown how tidal hydrodynamics in a salt marsh—tidal creek complex are linked to the morphology of the system. The link is established through the continuity equation and has been demonstrated through the use of the hypsometric curve and by relating the curve to the evolution of flood tidal currents.

Evaluation of the links requires integrated modeling and field studies that focus on the hydrology of the adjacent salt marsh and tidal creek complex. Models must successfully simulate flooding and drying of the salt marsh—tidal creek complex. No models should be used for predictive purposes until they are calibrated and validated.

Using some of the principles outlined in a review by Friedrichs and Perry (2001) and Pethick (2001), we suggest that the planning of land development taking place in and adjacent to tidal marshes take account of natural feedback between hydrodynamics and sediment supply. This means that (1) the heavy marsh grass density must be maintained to provide a favorable environment for suspended sediments to settle, (2) the natural sediment supply from land must be maintained and (3) there must be room for the marsh to migrate inland.

Acknowledgments

Our thanks to Brian Blanton at the University of North Carolina–Chapel Hill for help with the implementation of Equation 1 to the hypsometric data and to Anna Boyette at Skidaway Institute of Oceanography for preparation of the figures. We gratefully acknowledge support from the NOAA Center for Sponsored Coastal Ocean Research/Coastal Ocean Program, through the South Carolina Sea Grant Consortium pursuant to NOAA Award No. NA960PO113 and from a National Science Foundation Grant to the University of Georgia (LTER Grant No. OCE-9982133).

References

Bollinger, M. and W. Moore. 1993. Evaluation of salt marsh hydrology using radium as a tracer. *Geochimica Cosmochimica Aeta* 57:2203–2212.

Chalmers, A., R. Wiegert. and P. Wolf., 1985. Carbon balance in a salt marsh: interactions of diffusive export, tidal deposition and rainfall-caused erosion. *Estuarine Coastal and Shelf Science* 21:757–771.

Dronkers, J. 1986. Tidal asymmetry and estuarine morphology. *Netherlands Journal of Sea Research* 20:117–131.

Friedrichs, C. and D. Aubrey. 1988. Non-linear tidal distortion in shallow well-mixed estuaries: A synthesis. *Estuarine, Coastal and Shelf Science* 27:521-545.

Friedrichs, C. and J. Perry. 2001. Tidal salt marsh morphodynamics: A synthesis. *Journal of Coastal Research* SI 2 7:7–37.

Gardner, L. R. and M. Bohn. 1980. Geomorphic and hydraulic evolution of tidal creeks on a subsiding beach ridge plain, North Inlet (South Carolina). *Marine Geology* 34:91–97.

Gardner, L., W. K. Michener, B. Kjerfve, and D. Karinshakn. 1992. The geomorphic effects of Hurricane Hugo on an undeveloped coastal landscape at North Inlet, South Carolina. *Journal of Coastal Research* 8:181–186.

Healey, R., K. Pye, D. Stoddart, and T. Bayliss-Smith. 1981. Velocity variations in salt marsh tidal creeks. *Estuarine, Coastal and Shelf Science* 13:535–545.

Klinck, J., J. O'Brien, and H. Svendsen. 1981. A simple model of fjord and coastal circulation interaction. *Journal of Physical Oceanography* 11:1612-1626.

Leonard, L. A. and D. Reed. 2002. Hydrodynamics and sediment transport through tidal marsh canopies. *Journal of Coastal Research* SI 36:459–469.

Palmer, M. and G. Güst. 1985. Dispersal of meiofauna in a turbulent tidal creek. *Journal of Marine Research* 43:179–210.

Pethick, J. 2001. Coastal management and sea-level rise. *Catena* 42:307–322.

Postma, H. 1967. Sediment transport and sedimentation in the estuarine Environment, pp. 159–179. In, Lauff, G. (ed.), *Estuaries.* AAAS Publication No. 83, Washington, DC.

Schwing, F. and B. Kjerfve. 1980. Longitudinal characterization of a tidal marsh creek. *Estuaries* 3:236–241.

Schwing, F., L.-Y. Oey, and J. Blanton. 1988. Evidence for non-local forcing along the southeastern United States during a transitional wind regime. *Journal of Geophysical Research* 93:8221–8228.

Smith, J. and S. McLean. 1984. A model for flow in meandering streams. *Journal of Geophysical Research* 89:1301–1315.

Speer, P. and D. Aubrey. 1985. A study of non-linear tidal propagation in shallow inlet/estuarine systems, Part II: Theory. *Estuarine, Coastal and Shelf Science* 21:207–224.

Vörösmarty, and T. Loder III 1994. Spring-neap tidal contrasts and nutrient dynamics in a marsh-dominated estuary. *Estuaries* 17:537–551.

Wolaver, T., R. Dame, J. Spurrier, and A. Miller. 1988. Sediment exchange between a euhaline salt marsh in South Carolina and the adjacent tidal creek. *Journal of Coastal Research* 4:17–26.

Zarillo, G. 1985. Tidal dynamics and substrate response in a salt-marsh estuary. *Marine Geology* 67:13–35.

The Role of Tidal Wetlands in Estuarine Nutrient Cycling

Hank N. McKellar Jr. and Delma Bratvold

Summary by Hank N. McKellar Jr., Delma Bratvold, and Daniel R. Hitchcock

Excessive nutrient loading to coastal waters can cause intense algal blooms, oxygen depletion, and disruption of estuarine food webs. These impacts represent a threat to water quality and biological balances that maintain estuarine fisheries, recreation, and other beneficial uses of coastal waters. Recent trends in coastal development and population growth along the southeastern United States point to increases in nutrient loading from wastewater discharge and urban storm water runoff. To predict and manage the impacts of coastal development, these changes in nutrient loads need to be evaluated within the context of estuarine nutrient uptake, release, and cycling.

A major component of estuarine systems in the southeastern United States is an expansive area of tidal wetlands that surround a network of tidal creeks and channels. The daily tidal exchange and nutrient processing in these wetlands play an important role in regulating estuarine nutrient concentrations and water quality. While the primary pathway for nutrient movement between the larger estuary and the wetland is typically through a relatively predictable pattern of tidal water movement, the net import or export of nutrients from the wetland is also driven by the overall wetland metabolism, which is far more difficult to predict.

Overall wetland metabolism is the combined processes of growth, digestion, and energy production by all organisms in the wetland; it includes the processing of nutrients such as carbon, nitrogen, and phosphorus. Wetland metabolism is affected by nutrients brought to the wetland through the movement of tidal water, and upland surface water and groundwater. Within the wetland, organic matter is produced by photosynthetic plants and algae. This organic matter is digested by bacteria and other micro- and macrobiota found in both the surface water and sediment of the wetland. Nutrients are exchanged between the water and sediment. As floodwater ebbs over the intertidal regions, exposing the sediment surface to air, some of the floodwater is returned to the creeks over the marsh surface, while other floodwater seeps into the sediment and moves as groundwater toward the creek beds, where it enters the creek at low tide. This slower pathway for the return of tidal floodwater to the creek allows more time for the sediment biota to take up nutrients for energy and growth, and release metabolic products such as ammonia and carbon dioxide from the digestion of sediment organic matter.

A portion of the products of wetland metabolism is reused by other marsh organisms, causing a recycling of nutrients within the wetland. Some of the remainder may be exported from the wetland with ebb tides, and some may be sequestered in the wetland sediment. Substantial variations within and among wetlands in metabolic activity and recycling, sequestering, and tidal exchange are driven by factors such as geology, topography, hydrology, and associated sediment movement and accumulation, in addition to nutrient input from outside the wetland. This variety of contributing factors means that the same change in watershed contributions can have very different effects in different wetlands, and in the same wetland at different points in time.

In some cases, nutrient uptake by marshes may tend to buffer moderate impacts of nutrient loads from surrounding upland development. On the other hand, the natural export of organic debris from marsh grasses contributes to biochemical oxygen demand (BOD) in the estuarine water. This "natural" input should be accounted for in BOD waste load allocations to these productive systems. Wetland nutrient exchange may be particularly important in the estuaries of the southeastern United States, where a substantial tidal range (2–4 m) regularly flows across broad areas of productive intertidal marshes. These wetlands represent an important habitat where issues of coastal urban development and wetland function play a critical role in regulating estuarine water quality.

Although recent research has contributed much to our understanding of estuarine processes and how they affect nutrient exchanges in intertidal wetlands, more detailed understanding of marsh nutrient exchanges in urbanized estuaries is needed to address issues in land use planning and water quality management. Factors that affect both the uptake of nutrients by marshes and the potential export of matter from the marshes need to be better quantified to confidently predict and manage the effects of land use changes.

More effective water quality management plans may be able to be developed with greater consideration of a few key factors that affect the variation in uptake and release of nutrients from marshes (e.g., water movement and topography, successional stage, vegetation dominants). Application of these factors to management plans could be facilitated by the development of marsh categories based on these factors. These concepts need careful consideration in water quality management decisions for rapidly developing areas.

6.1 Introduction

Nutrient overenrichment of estuarine and nearshore waters is a problem common to an increasing number of coastal areas throughout the world (GESAMP 1990; Windom 1992; Galloway et al. 1996; Rabalais et al. 1996). It has been estimated that the present day anthropogenic input to ocean margins equals the natural input (GESAMP 1990; Galloway et al. 1996). Eutrophication of coastal waters is a potential threat to fisheries and recreational uses and poses a significant human health risk. Increased inputs of nutrients (particularly nitrogen and phosphorus) can have important consequences for the structure of estuarine communities, aquatic food webs, and estuarine water quality (Boynton et al. 1996; Rabalais et al. 1996). Excessive nutrient concentrations may also be responsible for stimulating harmful algal blooms, such as the toxic *Pfiesteria piscicida* blooms linked to fish kills in North Carolina, Chesapeake Bay, and Charleston Harbor (Burkholder et al. 1995; Lewitus et al. 1995).

The National Eutrophication Study (Bricker 1997) indicated clear reasons for concern with regard to eutrophication in many Atlantic and Gulf Coast estuaries of the United States. For the southeastern United States in particular, trends in both land use and population growth are major sources for concern. For example, the population of the coastal counties of Georgia and South Carolina has increased at a rate of about 20 percent each decade since about 1970 (Shabman 1996). This trend is expected to continue in this century. Along with population growth, coastal planners expect concurrent increases in wastewater discharge;

land cover changes to more urban land uses; and related impacts of nonpoint source runoff. All potentially lead to increased input of nutrients to coastal waters eutrophication (Livingston 1996; Stanley 1996).

Implications of increased nutrient loads need to be evaluated within the context of major estuarine processes of nutrient cycling and exchange. In particular, the daily tidal exchange and nutrient processing in estuarine wetlands plays an important role in regulating estuarine nutrient cycling and productivity (Odum 1968, 1980). The exchange between coastal wetlands and adjacent estuarine waters has been the subject of much research (Nixon 1980); recent summaries have described the emerging understanding of these exchanges (Childers et al. 2000). Wetland nutrient exchange may be particularly important in the estuaries of South Atlantic Bight (SAB), which are characterized by large, regular tides (2–4 m tidal range) that flow across broad areas of productive intertidal wetlands. With a typical ratio of intertidal wetland area to estuarine channel area of >3:1, southeastern U.S. estuaries, in particular, represent an important geographical area where issues of coastal urban development and wetland function play a critical role in regulating estuarine water quality. In this chapter, we examine some general concepts and recent developments in our understanding of wetland nutrient exchange. We also address key factors that may significantly affect these processes, such as surface sediment and groundwater; marsh age, successional stage, and vegetation dominants; organic matter content; temporal patterns; precipitation, land use, and nutrient loading.

6.2 Pathways of Wetland Exchange and Patterns of Nutrient Import/Export

In general, patterns of nutrient exchange between intertidal wetlands and estuarine waters indicate the dominant pathways of biogeochemical cycling within the wetland and the net influence of these wetlands on estuarine nutrients and water quality. Figure 6.1 illustrates several physical pathways of water and nutrient exchange that may result in patterns of net uptake or export. While tidal creek water quality is clearly influenced by water exchange with the ocean (1), important transformations occur along creek bottoms (2) and during tidal exchange with the intertidal wetlands (3,4). As floodtide waters inundate the vegetated marsh surface, nutrients may be removed from the overlying water (3) by a variety of mechanisms. Sediment trapping by the vegetation could lead to removal and burial of nutrients adsorbed onto the trapped particles

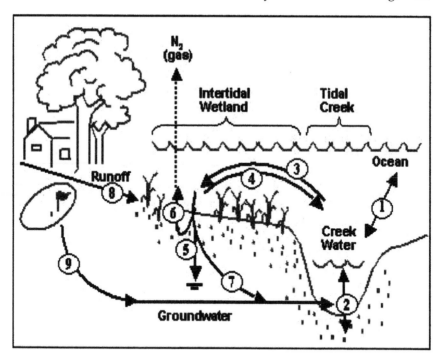

Fig. 6.1 Physical pathways of water and nutrient exchange in a tidal creek-wetland system. (1) Tidal water and associated nutrient exchange between the ocean and the wetlands. (2) Nutrient and organic matter transfer between the creek bed and creek water. (3) Uptake of tidal nutrients and organic matter by the marsh (surface vegetation and sediment). (4) Release of nutrients and organic matter by the marsh (surface vegetation and sediment). (5) Sedimentation and subsequent burial of tidal suspended solids (organic matter and associated nutrients) within the vegetated marsh. (6) Sediment nutrient cycling processes (e.g., nitrate assimilation into biomass, dissimilatory nitrate reduction to ammonia (DNRA), nitrification and denitrification). (7) Subsurface flow of interstitial water and associated organic matter and nutrients through the marsh sediment to the creek during low tide. (8) Upland surface water runoff. (9) Groundwater movement from upland sources to seepage.

(5). Some uptake by the marsh may be related to nutrient processing by the marsh vegetation (Morris 1980; Hopkinson and Shubauer 1984). However, microbial processes in the sediments, such as denitrification (6), dissimilatory nitrate reduction to ammonia (DNRA), and assimilation of nitrate into microbial biomass in the sediments and marsh litter, are also potential pathways of nutrient removal from the overlying water (Valiela and Teal 1979; Whiting and Morris 1986; Bowden et al. 1991; Rivera-Monroy and Twilley 1996).

The vegetated marsh surface may also release organic matter and nutrients to the tidal creek (4) in the form of detrital material, ranging in size from small particles to large mats of plant matter, produced by the marsh vegetation (Teal 1962; de la Cruz 1973). The washing of marsh surface organic matter and nutrients occurs both from daily tidal flow patterns and low-tide rainfall (Chalmers et al. 1985). Marsh sediment organic matter, derived from marsh grasses, algae, bacteria, and land plants (Pomeroy et al. 1977; Turner 1978; Klok, et al. 1984), releases both inorganic (Rocha 1998; Rocha and Cabral 1998) and dissolved organic nutrients (Goñi and Thomas 2000).

Some of the floodtide waters on the marsh surface may seep downward into the sediment pore waters, and it becomes entrained in subsurface flow back to the tidal creeks (7) during low tide drainage (Gardner 1975; Wolaver et al. 1980; and Whiting and Childers 1989). Finally, upland runoff (8) and upland groundwater (9) feed into the marsh system and may be affected by upland watershed land use (Valiela et al. 1992; Aelion et al. 1997; Wahl et al. 1997).

The balance among these complex interactions may culminate in a significant net movement of nutrients between the intertidal marshes and estuarine water. Early research using direct measures of tidal nutrient exchange in coastal marshes (Valiela and Teal 1979; Woodwell et al. 1979; Nixon 1980) suggested that intertidal salt marshes tend to export organic matter and some reduce inorganic nutrient fractions (ammonia) while importing suspended sediments and nitrate. More recent research has tended to corroborate the earlier work, although there is considerable variability in the reported values of net nutrient exchange for different marsh-estuarine systems.

6.3 Spatial Scale and Subsystem Interactions

While some studies of marsh nutrient exchange have relied on the use of relatively small-scale experimental chambers or flumes, others have based estimates of net nutrient exchange on larger-scale mass-flux measurements at the mouths of tidal creeks and channels. Results from these different spatial scales of investigation may differ considerably because of variable scales of subsystem interaction (Childers et al. 2000).

For example, marsh flume studies of nutrient flux across the *Spartina* marshes of North Inlet (South Carolina) found that the vegetated marsh surface imported nitrate + nitrite and ammonia, especially during the summer (Whiting et al. 1989). In contrast, transect sampling at the mouth of the estuary indicated that total marsh-estuarine ecosystem exported these nutrients (Whiting et al. 1987). The difference was due, in part, to the contribution of more estuarine subsystems (tidal creeks and subtidal benthos, and mudflats) and mechanisms (low-tide drainage, subtidal groundwater advection, and creek channel transformations) that culminated in nutrient export in the larger-scale study. Overall, considerations of spatial scale and subsystem interactions are essential for assessment of the relative significance of factors that affect wetland processes and their ultimate impact on the adjacent estuary.

6.4 Surface Sediment and Groundwater

The interactions of benthic and pelagic systems are well recognized as a critical component in estuarine nutrient cycling (e.g., Nixon 1980; Kelly and Nixon 1984; Mackin and Swider 1989; Hopkinson et al. 2001). Near-surface processes in marsh sediments have been suggested to be factors in the cycling of material in tidal estuaries (e.g., Goñi and Thomas 2000, and references therein). In a southeastern U.S. tidal estuary, both the proportion of organic carbon and organic nitrogen in the upper sediment have been shown to be lowest at lower elevations dominated by *Spartina*, higher at upper marsh elevations dominated by *Juncus*, and highest in the adjacent forest soil. Conversely, mineral content decreased from roughly 85 percent in *Spartina* regions, to 60 percent in *Juncus* regions, and finally to 20 percent in forest regions (Goñi and Thomas 2000). The presence of plant shoots induces lower surface water velocities and hence promotes sedimentation of suspended matter in the marsh (Leonard and Luther 1995). Small spatial-scale water currents and other sediment perturbations determine particle resuspension, and thus contribute to the extent to which local areas may be sources or sinks of suspended solids. It has been suggested that only a small percentage (1 to 5 percent) of *Spartina* production becomes part of the sediment (Middelburg et al. 1997). Mineral content of the upper sediment in the lower and upper marsh is largely composed of sand that is readily resuspended and deposited by tidal floodwater (Goñi and Thomas 2000). Consequently, in many tidal wetlands there is a net export of particulate matter from the surface sediment. Organic matter associated with the disturbed sediment is typically less dense than sand, and may be transported farther within a single tide.

In tidal wetlands, nutrient fluxes across the sediment-water interface are a function of both benthic remineralization and interstitial water advection, also referred to as seepage (Jordan and Correll 1985; Whiting and Childers 1989). Interstitial water that seeps into saltwater marshes and creeks has been reported to contain higher concentrations of nutrients than overlying water. Measurements of these interstitial water nutrients have included at least one or more of the following: ammonia, nitrate, dissolved organic nitrogen (DON), phosphates, and dissolved organic carbon (DOC) (Capone and Bautista 1985; Jordan and Correll 1985, Whiting and Childers 1989). In several South Carolina tidal creeks, interstitial water advection has been reported to provide several fold more ammonia and phosphate than low tide runoff from the marsh surface, while low tide surface runoff was a greater source of nitrate than interstitial seepage (Whiting and Childers 1989). Total groundwater input to coastal waters in the South Atlantic Bright has been estimated to be as much as 40 percent of the river water input (Moore 1996). The dominance of reduced inorganic nitrogen in low-oxygen interstitial water, and oxygenated inorganic nitrogen in surface runoff supports reports of oxygen depletion and anaerobic conditions in many marine sediments (e.g., Koike and Sørensen 1988; Mackin and Swider 1989; Cai and Sayles 1996).

Interstitial water seepage into tidal creeks is thought to be the result of hydraulic pressure generated by head differences between marsh pore water and the surface of the tidal creek water (Whiting and Childers 1989). Under some circumstances, the exchange of interstitial water and surface water may be promoted by changes in water density (Smetacek et al. 1976). In intertidal marsh regions with interstitial water that is of lower salinity than tidal inflow, flushing of the interstitial water into the surface water may be enhanced by gravity displacement of the lighter, lower-salinity water by the denser tidal water. These and other hydrological processes have been identified as contributing to groundwater seepage into marine systems (Burnett et al. 2003).

In addition to tidal water residence time, organic matter quality and quantity in the sediment affect the quantity of microbial metabolites such as CO_2, ammonia, DON, and PO_4^{3-} (Kelly and Nixon 1984; Christensen et al. 2000). The primary source of carbon in Winyah Bay, a river-dominated estuary in South Carolina, was found to be terrestrial, with increasing carbon from marine phytoplankton and, to a less extent, marsh grasses, toward the mouth of the largely salt marsh-surrounded bay (Goñi et al. 2003). However, within tidal marshes themselves, autochthonous carbon sources (i.e., marsh grasses, and, to a lesser extent, benthic algae) have been found to be the dominant carbon sources (Pomeroy et al. 1981; Goñi and Thomas 2000). In general, sediments with the highest amounts of biologically labile organic matter would be expected to have the greatest potential effect on nutrients as they move through the subsurface. This is a result of the greater biological activity and associated processing of carbon, nitrogen, and phosphorus that is common in sediments with more biologically degradable organic matter.

6.5 Marsh Age, Successional Stage, and Vegetation Dominants

In North Inlet, South Carolina, lower marsh regions dominated by the tall form of *S. alterniflora* were found to have higher nitrogen uptake and release rates compared to higher marsh areas dominated by short *S. alterniflora* (Whiting et al. 1989). Also in North Inlet, a younger, accreting marsh subsystem (Bly Creek) was found to import and accumulate more material than the total estuarine system of North Inlet, which includes large areas of more mature marsh that tended to export material (Dame et al. 1991). Similar patterns related to successional stages of wetland development were demonstrated in relation to marsh accretion and subsidence in the Gulf Coast marshes of Louisiana (Childers and Day 1990).

In many coastal systems, the successional stage of wetland development is further linked to marsh topography, tidal hydrology, and species dominants, all factors that probably have additional influence on net nutrient exchange in the wetland. Jordan and Correll (1991) demonstrated the effect of marsh elevation, frequency of tidal inundation, and dominant vegetation on nutrient exchange dynamics in the Rhode River estuary (Maryland). They found that the lower-elevation marsh (dominated by *Typha angustifolia*) tended to import nutrients via deposition of particulate matter, while the higher-elevation marsh (dominated by *Spartina patens, Distichilis spicata,* and *Scirpus olneyi*) tended to export nutrients (especially organic nitrogen and phosphorus).

Similar issues of marsh topography and vegetation dominants have been recently implicated in the tidal freshwater wetlands of the upper

Cooper River Estuary, South Carolina (McKellar et al. 2001; Morris et al. 2002; Huang and Morris 2003; Kelley et al. in prep). Many of these wetlands were managed for rice cultivation in the colonial times. Now these relict rice fields are mostly open to tidal exchange, but vary considerably in topography, tidal hydrology, vegetation dominants, and nutrient exchange. At lower elevations, the wetlands are charac-terized mostly by deeper, subtidal habitat dominated by submerged aquatic vegetation (*Hydrilla, Egeria,* and *Potamogeton*). At higher eleva-tions, the wetlands are characterized by intertidal habitats dominated by a diverse community of emergent marsh vegetation (Pickett et al. 1989), including *Zizaniopsis, Pontederia,* and *Peltandra.* These intertidal marsh sediments of the upper Cooper estuary generally contained higher organic content than the other successional stages (Huang and Morris 2003). Intermediate elevations are characterized by a combina-tion of subtidal and intertidal habitats dominated by vegetation with dense floating leaves and stems (*Ludwigia, Eichornia,* and *Polygonum*). Potential overall patterns of nitrate+nitrate (NO_x) uptake by all three wetland elevations are indicated by consistently lower NO_x concentra-tions in ebbing flow (Alford 2000; McKellar et al. 2001). However, the net uptake of NO_x was particularly consistent at the intertidal marsh, where flow-weighted ebb concentrations were usually 18–40 percent lower than the flow-weighted flood concentrations (Saroprayogi 2001).

These more consistent patterns of NO_x uptake by the intertidal marsh correlated with higher plant biomass and organic content of the sediments than in the other study sites (Huang and Morris 2003). Both of these factors may enhance NO_x uptake because of the increased organic substrate for denitrification (Nowicki et al. 1999) and DNRA (Christensen 2000). The coupling of nitrification and denitrification may also be stimulated by vegetative aeration of root-zone sediments (Reddy et al. 1989; Caffrey and Kemp 1992). Perhaps, the alternate wetting, drying, and aeration of surface sediments by tidal action in intertidal marshes further stimulate sediment denitrification and cor-responding NO_x uptake from the overlying water.

6.6 Organic Matter and Nitrogen Cycling

The net accumulation or export of organic matter in a particular region within a marsh results from the balance between local primary produc-tion and decomposition, in addition to resuspension and deposition as a result of surface water flow and groundwater seepage. Most of the primary production in tidal marshes is from marsh grasses (Pomeroy et al. 1981), and thus grasses provide the greatest component to the pool of particulate organic matter (POM). POM may be degraded to dissolved organic matter (DOM). As discussed by Alberts and Filip (1994), some river DOC is converted to particulate organic carbon (POC) as it mixes with seawater. While only 1–12 percent DOC was converted to POC in southeastern U.S. rivers, this represented a large change in POC, which is typically tenfold less than DOC in these systems. The distinction between DOC and POC is particle size, where as a practical matter, DOM is defined by a specific filter pore size through which it passes, commonly ranging from 0.2 to 0.7 μm.

The largest fraction of DOM in most natural waters is humic substances (HS), which are structurally complex organic matter (Thurman 1985). HS derive from both allochthonous vegetation drainage and autochthonous marsh grasses and algae. The variety of HS has contributed to varied reports of their utilization by bacterial flora, ranging from recalcitrant forms from which readily available energy sources have been exhausted (Rashid 1985), to labile, high-energy sources supporting up to 50 percent of the bacterial growth (Moran and Hodson 1990). In a survey of river and groundwater supplying five coastal U.S. estuaries, differences in organic matter composition explained 67–75 percent of the variation in bacterial growth (Hopkinson et al. 1998). In laboratory studies, microbial degradation of HS was found to be strongly dependent on HS composition and microorganism species (Hertkorn at al. 2002). The primary source of HS in tidal wetlands has been identified as marsh grasses, which may contribute roughly half of the HS on the southeastern U.S. continental shelf (Moran and Hodson 1994). Coastal marine HS, in particular, are used by bacterioplankton (Moran and Hodson 1994), and aerobic and anaerobic sediment bacteria (Alberts and Filip 1994; Coates et al. 1998).

In coastal environments, HS also provide an important direct source of nitrogen for bacteria, and an indirect source of nitrogen for primary producers as a result of bacterial nitrogen cycling (Carlsson et al. 1993). The dissolved organic nitrogen (DON) component of DOM represents the largest nitrogen pool in estuarine rivers of the Southeast (Alberts and Takács 1999), and may comprise up to 90 percent of the total dissolved nitrogen pool (i.e., dissolved inorganic and organic nitrogen) in southeastern estuarine waters (Alberts and Takács 1999; Bratvold, in prep). Thus, the combined quantity and lability of the DOM pool in tidal wetlands makes it an important consideration in carbon and nitrogen cycling.

Changes in the quantity of labile organic matter in sediments have been shown to correspond positively with rates of decomposition and associated nutrient cycling (Aller and Yingst 1980; Kelly and Nixon 1984; Marvin-DiPasquale and Capone 1998; Christensen 2000). In a stable isotope study of intertidal sediments, relatively labile organic matter of marine origin was preferentially degraded by the microbial community compared to the more recalcitrant terrestrial organic matter (Böttcher et al. 2000). The sorption of organic matter onto clay mineral surfaces was suggested to cause the correlation between organic matter burial and sediment clay content (Böttcher et al. 2000), which also corresponds to a lower rate of resuspension than sandy sediments (Middelburg et al. 2000).

The accumulation of organic matter in the marsh sediment may substantially reduce oxygen penetration and associated reduction-oxidation (i.e., redox) potential (Koike and Sørensen 1988; Cai and Reimers, 1995). Many nutrient cycling processes are sensitive to sediment redox potential. In coastal sediments with high organic content, nitrate concentrations fall rapidly with depth while ammonia concentrations increase (Koike and Sørensen 1988), indicating a more reduced environment. In contrast, coastal sediments with lower organic content typically exhibit a nitrate peak a few centimeters below the sediment surface as a result of nitrification, which occurs only in oxygenated sediment (Koike and Sørensen 1988). Tidal wetlands include sediments with a broad organic content range; thus, sediment depth profiles of nitrate and ammonia concentrations in different areas within tidal wetlands may fit each of these patterns, and may change seasonally, as discussed in section 6.7 (Temporal Patterns), below.

Redox potential and residence time are primary determinants of nutrient fluxes from marine groundwater (Slomp and Cappellen 2004). Furthermore, both carbon and nitrogen recycling increase substantially in the presence of active macrophytes and benthic microalgae, presumably due to their release of labile (high-quality) DOC that is rapidly used by the bacterial community (Middelburg et al. 1998). Thus, in tidal wetlands the prevalence of macrophytes and benthic microalgae may promote rapid cycling of carbon and nitrogen, the primary pathways of which may be determined by sediment reduction-oxidation potential.

The potential biogeochemical pathways for nitrogen cycling are shown in Figure 6.2. Nitrate may be assimilated into biomass both by autotrophs and heterotrophs (Caraco et al. 1998), or it may be reduced to either ammonia (via DNRA) or nitrogen gas (via denitrification). While some ammonia may be lost to the atmosphere, particularly under windy and high pH conditions, the products of denitrification are considered the primary means of nitrogen loss directly to the atmosphere.

Higher-sediment organics have been correlated with higher rates of both denitrification (Nowicki et al. 1999) and DNRA (Christensen 2000). In some tidal wetland regions, higher rates of denitrification may also be related to enhanced coupling of nitrification and denitrification (Kemp et al. 1990; Nowicki et al. 1999; LaMontagne et al. 2002) as a result of the cycling oxic conditions of the surface sediment due to alternate flooding and draining. However, in wetland regions with active benthic microalgal populations (Risgaard-Petersen 2003), nitrification rates may be very low due to high organic matter content (Strauss and Lamberti 2000), and anaerobic sediments (Focht and Verstraete 1977). Furthermore, denitrification is apparently inhibited by hydrogen sulfide (Sørensen et al. 1980), a byproduct of organic decomposition by sulfate-reducing bacteria under anaerobic, low-redox potential conditions. In shallow, temperate marine systems, organic content is positively

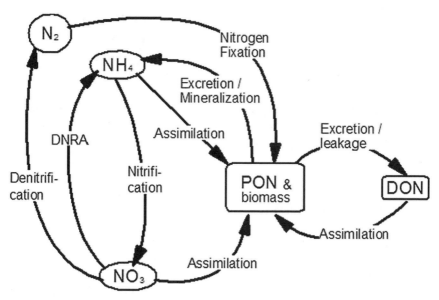

Fig. 6.2. Biogeochemical pathways of nitrogen cycling in tidal wetlands.

correlated with denitrification (Jensen et al. 1988). In subtropical tidal wetlands, denitrification was positively correlated with sediment organic content, but the proportion of nitrate processed by DNRA also increased with sediment organic content (Aelion in prep.).

Along coastal salinity gradients in Danish estuaries and the Rhône River outflow, the proportion of nitrate that is reduced via DNRA versus denitrification is greater in lower-salinity regions (Jørgensen and Sørensen 1985; Bianchi et al. 1994), which typically contain greater amounts of sediment organic matter. In the freshwater regions of both of these systems, as much as 90 percent of nitrate reduction was attributed to DNRA, while in the marine regions, the proportion of nitrate reduction attributed to DNRA ranged from 1–40 percent. DNRA is also important in anoxic marine sediments off the coast of Japan (Koike and Hattori 1978), where reduced rates of denitrification correlated with increased DNRA, decreased redox potential, reduced depth of oxygen penetration, and increased sediment organic matter (Koike and Sørensen 1988).

Thus, with respect to nitrate processing, the effects of increases in the organic load to a tidal wetland may vary depending upon the initial organic content and associated redox potential of the sediment. Modest organic loads to areas with relatively high redox may stimulate denitrification, but the same loads to areas with low redox may have a greater stimulatory effect on DNRA, increasing ammonia production and transport to the estuary. Overall, the vegetation and sediment variables throughout a tidal wetland likely cause substantial differences in the prominence of specific carbon- and nitrogen-processing pathways at different sites. While small spatial-scale measurements provide important information on system capability and variability, large spatial-scale measurements are better for assessment of the net contributions of tidal wetlands to broader estuarine systems.

6.7 Temporal Patterns

In addition to the daily tide and light cycles, temporal patterns of nutrient processing in tidal wetlands include spring and neap tide cycles and seasonal changes. Vörösmarty and Loder (1994) demonstrated significant changes in the nutrient chemistry of salt marsh tidal creeks as a function of marsh area inundated during spring and neap tides. During neap tides, relatively little area of the wetland was inundated. During spring tides, most of the wetland was inundated, and much lower concentrations of all nutrients (ammonium, phosphate, nitrate+nitrite, and total dissolved nitrogen and phosphorus) were observed in the estuarine waters suggesting a net nutrient uptake by the marsh. Mass balance computations for this system supported the hypothesis that significant levels of nutrient uptake by the intertidal wetlands were required to produce the observed spring/neap tide patterns.

Both oxygen uptake and sulfur reduction correlate with seasonal temperature (Hargrave and Phillips 1981; Jorgensen and Sorensen 1985; Mackin and Swider 1989; Böttcher et al. 2000), indicating temperature effects on both aerobic and anaerobic organic matter decomposition. The nitrogen cycle does not appear to be as greatly influenced by temperature as the carbon cycle (Jorgensen and Sorensen 1985). In the Chesapeake Bay, Bronk et al. (1998) assessed nitrogen cycling over three seasons and concluded that this system shifts from a primarily autotrophic system in the spring to an increasingly heterotrophic system in the summer and autumn. In particular, in the spring the Chesapeake Bay is driven by autotrophic uptake of riverine nitrate. As the seasons progress, increasingly more nitrogen is obtained from organic nitrogen provided by benthic and water column nitrogen regeneration. Correspondingly, ammonia and DON concentrations were greater in the autumn (Bronk et al. 1998). Greater autumn DON exports have also been reported in southeastern U.S. intertidal marshes (1690 ± 793 μ moles m^{-2} tide^{-1}), suggesting enhanced microbial decay and organic leaching from the accumulating litter and senescent vegetation at the end of the growing season (McKellar et al. 2001).

Processes such as microbial denitrification and DNRA in the intertidal sediments and nitrate assimilation by the marsh flora contribute to the net nitrate removal during tidal exchange. While the major source of nitrate for denitrification in most aquatic sediments appears to be produced within the sediments (Seitzinger 1988), uptake from the overlying water could represent a key process and may be seasonally important in some systems. In a shallow Danish estuary, nitrate was stable throughout the year except for a spring peak that was attributed to a peak in river nitrate input (Jørgensen and Sørensen 1985).

In the Chesapeake Bay, nitrification and denitrification were lowest in the late summer and highest in the spring and fall. These findings were correlated with sediment oxygen penetration and suggest a strong coupling between nitrification and denitrification, the former of which cannot function in anoxic conditions (Kemp et al. 1990). In contrast, in the North Inlet salt marshes (South Carolina), nitrate and ammonia uptake was higher in summer (Whiting et al. 1989). This nitrate pattern suggested high summer denitrification in this subtropical, intertidal system, where much of the surface sediment has daily exposure to air. This is further supported by measurements of the highest annual denitrification during the summer in other South Carolina tidal creeks, tributaries to the Okatee and Colleton Rivers (Joye and Aelion in prep.). Seasonal patterns of nitrate uptake by the intertidal community were also indicated in South Carolina rice field wetlands, where higher rates were seen during the spring and summer (>400 μmoles m^{-2} tide^{-1}) and lower rates during the fall and winter (<150 μmoles m^{-2} tide^{-1}) (McKellar et al. 2001).

6.8 Precipitation, Land Use, and Nutrient Loading

The relative importance of nutrient and sediment contributions from watershed runoff in particular may vary with rainfall and storm events. In a modeling study of the Hudson River estuary, precipitation significantly affected carbon flux to the estuary, and the magnitudes of day-to-day and seasonal precipitation patterns were more important than annual mean precipitation (Howarth et al. 1991). Whiting at al. (1989) found that wetland export increased during rain events because of disruption of marsh sediment, resulting from both storm intensity and the extent of exposed marsh sediment. The largest storm event in this study (2.25 cm over 1.28 h) caused a forty-fold increase in particulate nitrogen export, along with smaller increases in nitrate, ammonia, and DON (Whiting et al. 1989). Chalmers et al. (1985), described the cycling of POC in the tidal wetlands of Sapelo Island, Georgia, where POC was deposited in marsh surfaces during daily tidal exchanges, and later exported during rain events. A moderate rain event in the Okatee estuary of South Carolina (i.e., 4.75 cm over about 2 hours) caused a two orders of magnitude increase in total suspended solids (TSS) (Bratvold, in prep) at a site midway between the creek headwaters and mouth, suggesting substantial movement of sediment.

Sediment resuspension in the Laguna Madre estuary (Texas) results in roughly equal amounts of ammonia release from three compartments: porewater, ammonia loosely bound to the sediment, and ammonia tightly bound to the sediment (Morin and Morse 1999). This suggests that the magnitude of an ammonia pulse as a result of rain event sediment resuspension may depend upon both sediment nutrient content and the geochemical substrate (e.g., sand and clay). Overall, the magnitude of nutrient fluxes during moderate and large precipitation events and the resulting response of the intertidal marsh is a potentially important topic that has had relatively little address.

Changes in land use patterns, such as increases in impervious surfaces and retention ponds, may cause less-frequent, but greater, single-event releases of nutrients from upland areas. The modeling study of the Hudson River estuary (Howarth et al. 1991) suggested that urban, suburban, and agricultural areas provide substantially more sediment and organic carbon inputs to the estuary than forested areas.

Issues of upland development, urban runoff, and wastewater discharge into estuarine systems are particularly important in evaluating trends toward estuarine eutrophication and the corresponding roles of tidal wetlands. In highly urbanized/industrialized estuaries (such as the Savannah River and Charleston Harbor) nutrient inputs tend to be dominated by wastewater discharges and permit enforcement (Windom et al. 1998). In the absence of major point sources of wastewater discharge, upland commercial and residential development can still lead to elevated nutrient loads due to nonpoint source runoff. Wahl et al. (1997) found that estuarine nitrogen loading from a residential/commercialized coastal watershed (Murrell's Inlet, South Carolina) was two times higher than from a nearby, forested watershed. The higher loading resulted, in part, from higher nitrate concentrations in the stream water, due to higher concentrations in the shallow groundwater (Aelion et al. 1997). Additional nitrate contributions have been associated with surface runoff from impervious surfaces (Wahl et al. 1997). Furthermore, the urban runoff was characterized by higher rates of stream discharge, due largely to channelization of the urban streams.

The corresponding estuarine impacts of upland nutrient loading must be examined in the context of estuarine nutrient processing. In particular, we need to understand how the processes of tidal nutrient exchange in the impacted wetlands respond to increased loading and the resulting higher background nutrient concentrations. In the relatively eutrophic waters of the Patuxent River (Cheseapeake Bay tributary, Maryland), Merrill and Cornwell (2000) found a loose

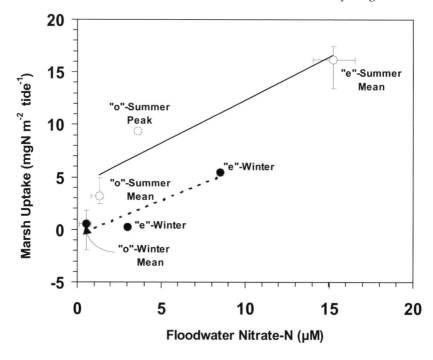

Fig. 6.3. Comparison of nitrate uptake rates as functions of floodwater nitrate concentration and season. Data are from two marsh sites in South Carolina that exhibited widely different ambient levels of nitrate-N. "o-" in the data labels indicates data from an oligotrophic marsh (Bly Creek, North Inlet, from Whiting et al. 1989). "e-" indicates data from a eutrophic marsh (Goose Creek, Cooper River estuary, from McKellar et al. 1996). Open circles as data points represent summer data (water temperature 25–30°C), filled circles represent winter data (water temperature 10–15°C). Crosses indicate the ranges of seasonal observations. The solid line and dashed line represent a linear regression through the summer and winter data, respectively.

positive correlation between rates of denitrification in the marsh sediments and the ambient nitrate concentration, with higher rates occurring during summer peaks in concentration. In South Carolina, nitrate uptake by intertidal marshes (i.e., combined nitrate assimilation, denitrification, and DNRA) has been examined across a moderate range of ambient nitrate concentrations. In the low-nutrient marshes of North Inlet (South Carolina), there is relatively little upland development or wastewater discharge. Tidal floodwater concentrations of

1991. Annual material processing by a salt marsh-estuarine basin in South Carolina, USA. *Marine Ecology Progress Series* 72:153–166.

de la Cruz, A. A. 1973. The role of tidal marshes in the productivity of coastal waters. *Association of Southeastern Biologists Bulletin* 20:147–156.

Focht, D. D. and W. Verstraete. 1977. Biochemical ecology of nitrification and denitrification. *Advances in Microbial Ecology* 1:135–214.

Galloway, J. N., R. W. Howarth, A. F. Michaels, S. W. Nixon, J.M.Prospero, and F. J. Dentener. 1996. Nitrogen and phosphorus budgets of the North Atlantic Ocean and its watershed. *Biogeochemistry* 35:3–35.

Gardner, L. R. 1975. Runoff from an intertidal marsh during low tide exposure. Recession curves and chemical characteristics. *Limnology and Oceanography* 20:81–89.

GESAMP. 1990. State of the Marine Environment. Reports and Studies No. 39. Joint Group of Experts on the Scientific Aspects of Marine Pollution. Nairobi, United Nations Environment Programme. 111 p.

Goñi, M. A. and K. A. Thomas. 2000. Sources and transformations of organic matter in surface soils and sediments from a tidal estuary, North Inlet, South Carolina, USA. *Estuaries* 23:548–564.

Goñi, M. A., M. J. Teixeira, and D. W. Perkey. 2003. Sources and distribution of organic matter in a river dominated estuary (Winyah Bay, SC, USA). *Estuarine Coastal and Shelf Science* 57:1023–1048.

Hargrave, B. T. and G. A. Phillips. 1981. Annual in situ carbon dioxide and oxygen flux across a subtidal marine sediment. *Estuarine Coastal and Shelf Science* 12:725–737.

Hertkorn, N., H. Claus, P. H. Schmitt-Kopplin, E. M. Perdue, and Z. Filip. 2002. Utilization and transformation of aquatic humic substances by autochthonous microorganisms. *Environmental Science and Technology* 36:4334–4345.

Hopkinson, C. S., I. Buffam, J. Hobbie, J. Vallino, M. Perdue, B. Eversmeyer, F. Prahl, J. Covert, R. Hodson, M. A. Moran, E. Smith, J. Baross, B. Crump, S. Findlay, and K. Foreman. 1998. Terrestrial inputs of organic matter to coastal ecosystems: an intercomparison of chemical characteristics and bioavailability. *Biogeochemistry* 43:211–234.

Hopkinson, C. S., A. E. Giblin, and J. Tucker. 2001. Benthic metabolism and nutrient regeneration on the continental shelf of Eastern Massachusetts, USA. *Marine Ecology Progress Series* 224:1–19.

Hopkinson, C. S. and J. P. Schubauer. 1984. Static and dynamic aspects of nitrogen cycling in the salt marsh graminoid Spartina alterniflora. *Ecology* 65:961–969.

Howarth, R. W., J. R. Fruci, and D. Sherman. 1991. Inputs of sediment and carbon to an estuary ecosystem: Influence of land-use. *Ecological Applications* 1:27–39.

Huang, X. and J. Morris. 2003. Trends in phosphatase activity along a successional grandient of tidal freshwater marshes on the Cooper River, South Carolina. *Estuaries* 26:1281–1290.

Jensen, M. H., T. K. Andersen, and J. Sørensen, 1988. Denitrification in coastal bay sediment: regional and seasonal variation in Aarhus Bight, Denmark. *Marine Ecology Progress Series* 48:155–162.

Jordan, T. E. and D. L. Correll. 1991. Continuous automated sampling of tidal exchanges of nutrients by brackish marshes. *Estuarine, Coastal and Shelf Science* 32:527–545.

Jordan, T. E. and D. L. Correll. 1985. Nutrient chemistry and hydrology of interstitial water in brackish tidal marshes of Chesapeake Bay. *Estuarine, Coastal and Shelf Science* 21:45–55.

Jørgensen, B. B. and J. Sørensen. 1985. Season cycles of O_2, NO_3^- and SO_4^{2-} reduction in estuarine sediments: the significance of an NO_3^- reduction maximum in spring. *Marine Ecology Progress Series* 24:65–74.

Joye, S. 2004. Groundwater and sediment biogeochemistry in the Okatee River Estuary, p. 106–110. In, South Atlantic Bight Land Use-Coastal Ecosystem Study (LU-CES) Phase II Final Progress Report, South Carolina Sea Grant Consortium, Charleston, SC.

Kemp, M., P. Sampou, J. Caffrey, M. Mayer, K. Henriksen, and W. Boynton. 1990. Ammonium recycling versus denitrification in Chesapeake Bay sediments. *Limnology and Oceanography* 35:1545–1563.

Kelley, B. J., D. L. Tufford, J. T. Morris, and L. Hardison. (In preparation.) Macrophyte community succession in former rice fields on the Cooper River, South Carolina.

Kelly, J. R. and S. Nixon. 1984. Experimental studies of the effect of organic deposition on the metabolism of a coastal marine bottom community. *Marine Ecology Progress Series* 17:157–169.

Klok, J., M. Baas, H. C. Cox, J. W. DeLeeuw, W. I. C. Rijpstra, and P. A. Schenck. 1984. Qualitative and quantitative characterization of the total organic matter in a recent marine sediment (Part II). *Organic Geochemistry* 6:265–279.

Koike, I. and A. Hattori. 1978. Denitrification and ammonia formation in anaerobic coastal sediments. *Applied Environmental Microbiology* 35:278–282.

Koike, I. and J. Sørensen. 1988. Nitrate reduction and denitrification in marine sediments, pp. 251–273. In, Blackburn, T. H. and J. Sørensen (eds.), Nitrogen Cycling in Coastal Marine Environments. John Wiley and Sons, New York, NY.

LaMontagne, M., V. Astorga, A. Giblin, and I. Valiela. 2002. Denitrification and the stoichiometry of nutrient regeneration in Waquoit Bay, Massachusetts. *Estuaries* 25:272–281.

Leonard, L. A. and M. E. Luther. 1995. Flow hydrodynamics in tidal marsh canopies. *Limnology and Oceanography* 40:1474–1484.

Lewitus, A. J., R. V. Jesien, T. M. Kana, J. M. Burkholder, H. B. Glasgow Jr., and E. May. 1995. Discovery of the "phantom" dinoflagellate in Chesapeake Bay. *Estuaries* 18:373–378.

Livingston, R. 1996. Eutrophication in estuaries and coastal systems: Relationships of physical alterations, salinity stratification, and hypoxia, pp. 285–318. In, Vernberg, J., W. Vernberg, and T. Siewicki (eds.), Sustainable Development in the Southeastern Coastal Zone. University of South Carolina Press, Columbia.

Mackin, J. E. and K. T. Swider. 1989. Organic matter decomposition pathways and oxygen consumption in coastal marine sediments. *Journal of Marine Research* 47:681–716.

Marvin-DiPasquale, M. C. and D. G. Capone. 1998. Benthic sulfate reduction along the Chesapeake Bay central channel. I. Spatial trends and controls. *Marine Ecology Progress Series* 168:213–228.

McKellar, H., A. Smith, A. Douglass, and R. Rao. 1995. Wetland nutrient exchange in an urbanized estuary: Relationships to point-source Discharges and nonpoint runoff, p. 85. Program Abstracts for the 13th International Estuarine Research Federation Conference. Estuarine Research Foundation, Port Republic, MD.

McKellar, H., P. Saroprayogi, M. Alford, J. Kelley, and J. Morris. 2001. Tidal nutrient fluxes in relict rice field wetlands: Relations to vegetation dominants and succession. Invited oral presentation at the 16th International Biennial Estuarine Research Federation Conference, September, 2001, St. Petersburg, FL.

Merrill, J. and J. Cornwell. 2000. The role of oligohaline marshes in estuarine nutrient cycling, pp. 425–441. In, Weinstein, M. and D. Kreeger (eds.), Concepts and Controversies in Tidal Marsh Ecology. Kluwer Academic Publishers, Boston, MA.

Middelboe, M, N. Kroer, N. O. G. Jørgensen, and D. Pakulski. 1998. Influence of sediment on pelagic carbon and nitrogen turnover in a shallow Danish estuary. *Aquatic Microbial Ecology* 14:81–90.

Middelburg, J. J., C. Barranguet, H. T. S. Boschker, P. M. J. Herman, T. Moens, and C. H. R. Heip. 2000. The fate of intertidal microphytobenthois carbon: An in situ 13C-labeling study. *Limnology and Oceanography* 45:1224–1234.

Middelburg, J. J., J. Nieuwenhuize, R. K. Lubberts, and O. van de Plassche. 1997. Organic carbon isotope systematics of coastal marshes. *Estuarine, Coastal and Shelf Science* 45:681–687.

Moore, W. S. 1996. Large groundwater inputs to coastal waters revealed by [226]Ra enrichments. *Nature* 380:612–614.

Moran, M. A. and R. E. Hodson. 1990. Bacterial production on humic and nonhumic components of dissolved organic carbon. *Limnology and Oceanography* 35:1744–1756.

Moran, M. A. and R. E. Hodson. 1994. Dissolved humic substances of vascular plant origin in a coastal marine environment. *Limnology and Oceanography* 39:762–771.

Morin, J. and J. W. Morse. 1999. Ammonium release from resuspended sediments in the Laguna Madre estuary. *Marine Chemistry* 65:97–110.

Morris, J. T. 1980. The nitrogen uptake kinetics of *Spartina alterniflora* in culture. *Ecology* 61:1114–1121.

Morris, J., J. Bulak, J. Kelley, and H. McKellar. 2002. Structure and Functions of tidal freshwater wetlands on the Cooper River, SC: Effects of water management on succession, nutrient cycling, and fish habitat. Annual Progress Report, South Carolina Sea Grant Consortium, Charleston, SC.

Nixon, S. 1980. Between coastal marshes and coastal waters: A review of twenty years of speculation and research on the role of salt marshes in estuarine productivity and water chemistry, pp. 437–525. In, Hamilton, P. and K. B. Macdonald (eds.), Estuarine and Wetland Processes. Plenum, New York, NY.

Nowicki, B., E. Requintina, D. Keuren, and J. Portnoy. 1999. The role of sediment denitrification in reducing groundwater-derived nitrate inputs to Nauset marsh estuary, Cape Cod, Massachusetts. *Estuaries* 22:245–259.

Odum, E. P. 1968. A research challenge: Evaluating the productivity of coastal and estuarine water, pp. 63–64. In, Proceedings of the Second Sea Grant Conference. University of Rhode Island, Newport, RI.

Odum, E. P. 1980. The status of three ecosystem-level hypotheses regarding salt marsh estuaries: tidal subsidy, outwelling and detritus-based food chains, pp. 485–495. In, Kennedy, V. (ed.), Estuarine Perspectives. Academic Press, New York, NY.

Pickett, J., H. McKellar, and J. Kelley. 1989. Community composition, leaf mortality, and net primary production in a tidal freshwater marsh in South Carolina, pp. 351–364. In, Sharitz, R. R. (ed.), Freshwater Wetlands and Wildlife. Department of Energy Symposium Series No. 61, U.S. DOE Office of Science and Technology Information, Oak Ridge, TN.

Pomeroy, L., K Bancroft, J. Breed, R. Christian, D. Frankenberg, J. Hall, L. Maurer, W. Wiebe, R. Wiegert, and R. Wetzel. 1977. Flux of organic matter through a salt marsh, pp. 270–279. In, Wiley, M. (ed.), Estuarine Processes, Vol. 2. Academic Press, New York, NY.

Pomeroy, L. R., W. M. Darley, E. L. Dunn, J. L. Gallagher, E. B. Haines, D. M. Whitney. 1981. Primary production, pp. 39–68. In, Pomeroy, L. R. and R. G. Weigert (eds.), The Ecology of a Salt Marsh. Springer, Berlin.

Rabalais, N. N., R. E. Turner, Q. Dortch. W. J. Wiseman Jr., and B. K. Sen Gupta. 1996. Nutrient changes in the Mississippi River and system responses on the adjacent continental shelf. *Estuaries* 19:386–407.

Rashid, M. A. 1985. Geochemistry of Marine Humic Compounds. Springer-Verlag, New York, NY. 300 p.

Reddy, K. R., W. H. Patrick, and C. W. Lindau. 1989. Nitrification-denitrification at the plant root-sediment interface in wetlands. *Limnology and Oceanography* 34: 1004–1013.

Risgaard-Petersen, N. 2003. Coupled nitrification-denitrification in autotrophic and heterotrophic estuarine sediments: On the influence of benthic microalgae. *Limnology and Oceanography* 48:93–105.

Rocha, C. 1998. Rhythmic ammonium regeneration and flushing in intertidal sediments of the Sado estuary. *Limnology and Oceanography* 43:811–822.

Rocha, C. and A. P. Cabral. 1998. The influence of tidal action on porewater nitrate concentration and dynamics in intertidal sediments of the Sado estuary. *Estuaries* 21:635–645.

Rivera-Monroy, V. H. and R. R. Twilley. 1996. The relative role of denitrification and immobilization in the fate of inorganic nitrogen in mangrove sediments (Termino Lagoon, Mexico). *Limnology and Oceanography* 41:271–283.

Saroprayogi, P. 2001. Tidal exchange on nutrients in the freshwater wetlands on the Upper Cooper River, SC. Master's thesis, University of South Carolina, Columbia. 69 p.

Seitzinger, S. 1988. Denitrification in freshwater and coastal marine ecosystems: Ecological and geochemical significance. *Limnology and Oceanography* 33:702–724.

Shabman, L. 1996. Land settlement, public policy, and the environmental future of the Southeast coast, pp. 7–24. In, Vernberg, J., W. Vernberg, and T. Siewicki (eds.), Sustainable Development in the Southeastern Coastal Zone. University of South Carolina Press, Columbia.

Slomp, C. P. and P. Van Cappellen, 2004. Nutrient inputs to the coastal ocean through submarine groundwater discharge: controls and potential impact. *Journal of Hydrology* 295:64–86.

Smetacek, V., B. von Bodugen, K. von Bröckel, and B. Zeitzschel. 1976. The plankton tower. II. Release of nutrients from sediments due to changes in the density of bottom water. *Marine Biology* 34:373–378.

Sørensen, J., J. M. Tiedje, and R. B. Firestone. 1980. Inhibition of sulfide by nitric of nitrous oxide reduction by denitrifying *Pseudomonas fluorescens*. *Applied Environmental Microbiology* 39:105–108.

Stanley, D. 1996. Long-term trends in nutrient generation by point and nonpoint sources in the Albemarle-Pamlico estuarine basin, pp. 319–342. In, Vernberg, J., W. Vernberg, and T. Siewicki (eds.), Sustainable Development in the Southeastern Coastal Zone. University of South Carolina Press, Columbia.

Strauss, E. A., and G. A. Lamberti. 2000. Regulation of nitrification in aquatic sediments by organic carbon. *Limnology and Oceanography* 45:1854–1859.

Teal, J. M. 1962. Energy flow in the salt marsh ecosystem of Georgia. *Ecology* 43:614–624.

Thurman, E. M. 1985. Organic geochemistry of natural waters. Martinus Nijhoff/Dr W. Junk, Boston, MA.

Turner, R. 1978. Community plankton respiration in a salt marsh estuary and the importance of macrophyte leachates. *Limnology and Oceanography* 23:442–451.

Valiela, I., K. Foreman, M. LaMontagne, D. Hersh, J. Costa, P. Peckol, B. DeMeo-Anderson, C. D'Avanzo, M. Babione, C. H. Sham, J. Brawley, and K. Lajtha. 1992. Coupling of watersheds and coastal waters: sources and consequences of nutrient enrichment in Waquoit Bay, Massachusetts. *Estuaries* 15:443–457.

Valiela, I. and J. M. Teal. 1979. The nitrogen budget of a salt marsh ecosystem. *Nature* 280:652–656.

Vörösmarty, C. J. and T. C. Loder 1994. Spring-neap tidal contrasts and nutrient dynamics in a marsh-dominated estuary. *Estuaries* 17:537–551.

Wahl, M., H. McKellar, and T. Williams. 1997. Patterns of nutrient loading in forested and urbanized coastal streams. *Journal of Experimental Marine Biology and Ecology* 213:111–132.

Whiting, G. and D. Childers. 1989. Subtidal advective water flux as a potentially important nutrient input to southeastern USA salt-marsh estuaries. *Estuarine, Coastal and Shelf Science* 28:417–431.

Whiting, G. J., H. N. McKellar Jr., B. Kjerfve, and J. Spurrier. 1987. Tidal exchange of nitrogen between a southeastern salt marsh and the coastal ocean. *Marine Biology* 95:173–182.

Whiting, G. J., H. N. McKellar Jr., and T. Wolaver. 1989. Nitrogen exchange between a portion of vegetated salt marsh and the adjoining creek. *Limnology and Oceanography* 32:463–473.

Whiting, G. and J. Morris. 1986. Nitrogen fixation in a salt marsh: its relationship to temperature and an evaluation of an in situ chamber technique. *Soil Biology and Biochemistry* 18:515–521.

Windom, H. L. 1992. Contamination of the marine environment from land-based sources. *Marine Pollution Bulletin* 25:1–4.

Windom, H. L., H. McKellar, C. Alexander, A. Craig, A. Abusam, and M. Alford. 1998. Indicators of trends towards coastal eutrophication. Final Report to South Carolina Sea Grant, LU-CES Program, Charleston, SC. 68 p. http://www.lu-ces.org/documents/SOKreports/coastaleutrophication.pdf.

Wolaver, T. G., L. Wetzel, J. Zieman, and K. Webb. 1980. Nutrient interactions between salt marsh, mudflats, and estuarine water, pp. 123–134. In, Kennedy, V. S. (ed.), Estuarine Perspectives. Academic Press, New York, NY.

Woodwell, G. M., C. A. S. Hall, D. E. Whitney, and R. A. Houghton. 1979. The Flax Pond ecosystem study: Exchanges of inorganic nitrogen between an estuarine marsh and Long Island Sound. *Ecology* 60:695–702

7

Evaluating the Potential Importance of Groundwater-Derived Carbon, Nitrogen, and Phosphorus Inputs to South Carolina and Georgia Coastal Ecosystems

Samantha B. Joye, Deborah A. Bronk, Dirk J. Koopmans, and Willard S. Moore

Summary by Samantha B. Joye and Daniel R. Hitchcock

Groundwater serves as a critical source of fresh, potable water for humans and is also an important water source for agriculture, both for cropland irrigation and livestock watering. Concern about groundwater resources has increased substantially in recent years because both the quality and quantity of groundwater resources have been degraded in many areas. It is likely that groundwater resources will be affected by changing land use patterns in the coastal Southeast. Management of groundwater resources and amelioration of impacts due to development in Georgia and South Carolina will depend on the availability of a comprehensive understanding of groundwater chemistry and the ability to meet basic research needs related to groundwater dynamics.

Groundwater collectively refers to the water contained within the empty spaces between rocks in the soil subsurface. Much of the rain falling on land is stored for at least some period of time in the groundwater reservoir. Groundwater is an important source of freshwater to streams, rivers, lakes, and even to coastal estuaries. Aquifers are permeable layers in the subsurface through which groundwater flows easily and somewhat rapidly. Water inputs to aquifers occur in areas of recharge, while water removal from aquifers occurs in areas of discharge. Recharge occurs via the percolation of water into the porous aquifer, while discharge occurs naturally in artesian springs or wells. There are two types of aquifers, shallow and deep. Shallow aquifers lie within a few meters of the soil surface and may be constantly recharged with water seeping into the aquifer all along the flow path. Deep aquifers lie tens or more meters below the soil surface and are recharged in localized areas that are typically far from discharge points.

Water flow through aquifers is driven by pressure gradients that occur on a variety of spatial scales, but, naturally, water always flows downhill. On small spatial scales, groundwater flows from high to low elevation, meaning that areas of low elevation, such as stream or river beds, have high groundwater inputs, while high-elevation areas, such as hills or mountain tops, have no groundwater inputs (in fact, these may be areas of recharge). On larger spatial scales, groundwater flows generally from inland areas of high elevation to near-shore regions of lower elevation.

Groundwater in shallow aquifers is particularly susceptible to contamination because it is constantly recharged along (often) long flow paths. Therefore, activities that occur within the recharge zone of shallow aquifers can potentially impact the chemical signature of groundwater. Surficial groundwater can be thought of as a shallow water circulatory system that connects regions of differing land use. In areas of agricultural land use, fertilizer-derived nutrients, such as nitrogen and phosphorus, as well as pesticides and herbicides, can become enriched in groundwater. Similarly, urbanization can alter patterns of groundwater recharge by increasing the area of impermeable surface (e.g., as a result of paving or home building) as well as by enriching groundwater with fertilizers, pesticides, herbicides, and petroleum products.

Nutrients and other chemicals are passively transported along with groundwater through watersheds and may be delivered to surficial water systems in areas of passive discharge or where groundwater is actively pumped. Currently the fraction of the groundwater chemical load delivered via natural discharge is unknown. Concern about groundwater-derived chemical inputs is particularly relevant in coastal ecosystems, where changing inputs of such nutrients as nitrogen and phosphorus can alter the patterns and types of primary production. Few studies in the southeastern United States have examined the nutrient content of groundwater. At present, the data necessary to evaluate the role of groundwater as a source of nutrients to coastal waters and to assess the impact of land use on groundwater biogeochemistry and materials fluxes are only beginning to accumulate.

Serious declines in coastal water quality and ecosystem health have resulted from population growth and agricultural, commercial, and industrial activities in coastal watersheds, as well as from increased loading of anthropogenic wastes (organics and nutrients) originating at localized (e.g., wastewater treatment plants, factories) and diffuse (e.g., urban and agricultural runoff) sources. Ultimately, when excessive amounts of anthropogenic materials arrive in coastal waters, they often lead to eutrophication, which can be loosely defined as the increase of labile organic matter supplies to an ecosystem. Visible signs of eutrophication, including increased frequency of harmful algal blooms, water column hypoxia or anoxia (i.e., partial or complete depletion of the oxygen dissolved in the water), fish kills, and generally reduced water quality, are apparent in coastal environments across the globe. Understanding the causes of eutrophication and documenting ecosystem responses to eutrophication are key research challenges facing coastal ecologists today.

Groundwater is an important, but poorly understood, source of nutrients and organic materials to coastal waters. Along the coast of Georgia and South Carolina, groundwater inputs are thought to be important, but the input terms have not been quantified in most areas. The lack of information regarding groundwater flux and groundwater quality makes it impossible to predict how the quality (chemistry) or quantity (freshwater flux) of this "source term" will respond to increased development pressures in coastal regions.

A comparison of available data on the importance of surface (river, stream) and groundwater (confined subterranean and unconfined water table aquifers) sources of carbon and nutrients to South Carolina and Georgia coastal systems indicates that the data are currently insufficient to estimate groundwater-derived nutrient fluxes. It is possible, however, to identify the data gaps. Research needed to describe groundwater biogeochemistry and to quantify groundwater fluxes, and the linkages between groundwater nutrient content and land use is both feasible and critical to the management of water supplies in the rapidly developing coastal Southeast.

7.1 Introduction

As human population has grown, agricultural, commercial, and industrial activities have expanded in coastal watersheds. These activities have increased inputs of anthropogenic wastes from localized (e.g., sewage, industrial effluent) and diffuse (e.g., agricultural runoff) sources to surface waters and groundwater (Westerman et al. 1995; Jordan et al. 1997a, b). Ultimately, increased nutrient loading to coastal waters through these and other sources causes eutrophication, the increase in labile organic matter supply to an ecosystem (Nixon 1995). The recent increases in frequency of harmful algal blooms, water column hypoxia and anoxia, fish kills, and reduced water quality (Nixon 1995; Paerl et al. 1998) is likely due in total or in part to eutrophication.

A variety of factors influence surface water quality, including watershed land use (e.g., natural vs. developed), type of development (e.g., residential, agricultural, or commercial), soil type and erosion rate (Hill 1978; Correll 1981; Beaulac and Reckhow 1982; Walling and Webb 1985; Frink 1991; Correll et al. 1992; Nearing et al. 1993; Jordan and Weller 1996; Jordan et al. 1997a). Significant effort has focused on documenting the chemical composition (quality) and flow (quantity) of surface waters (e.g., rivers, streams), which make up the most conspicuous component of the hydrologic cycle (Ittekkot 1988; Frink 1991; Humborg et al. 1997). However, accurate models of the response of coastal ecosystems to land use change require the inclusion of all relevant sources of nutrients and organic materials. Concentrations of nutrients and organic matter in groundwater frequently exceed those in surface waters (Sewell 1982 and others, see below), suggesting that groundwater is a potentially important source of these materials to coastal waters. Unfortunately, groundwater-derived fluxes have not been quantified in most coastal environments. The current lack of information regarding groundwater flux and groundwater quality in coastal systems makes it impossible to predict how the quality (chemical composition) or quantity (input rate or flux) of this source term will respond to increased development pressures in upstream watersheds and within coastal regions.

Spatial and temporal variations in groundwater quality and quantity determine its potential role in influencing coastal ecosystem productivity and possibly eutrophication. Existing data suggest that the input of fresh or salty groundwater to coastal systems is equivalent to between 5 and 40 percent of surface water inputs (Cable et al. 1996, 1997; Moore 1996; Hussain et al. 1999; Li et al. 1999). Because nutrient concentrations in groundwater often exceed those in surface waters by an order of magnitude or more, groundwater-associated nutrient fluxes may significantly impact ecosystem-level nutrient budgets (Valiela et al. 1978, 1990; Sewell 1982; Capone and Bautista 1985; LaPointe and O'Connell 1989; Capone and Slater 1990; Giblin and Gaines 1990; LaPointe et al. 1990; Portnoy et al. 1998; Corbett et al. 1999). Groundwater discharge has been documented as an important source of nutrients and metals in some coastal ecosystems (Johannes 1980; Moore 1996, 1999; Shaw et al. 1998; Krest et al. 1999). However, only a few data sets contain estimates of both the flux and chemical composition of groundwater (e.g., Moore 1996; Portnoy et al. 1998; Corbett et al. 1999; Krest et al. 2000), making it difficult to constrain groundwater-derived nutrient loading rates. Studies linking groundwater-derived nutrient inputs to the biological response of a coastal ecosystem (e.g., as a nutrient source for primary producers or as an organic source for heterotrophic bacteria) exist for only a handful of environments (LaPointe and O'Connell 1989; Capone and Slater 1990; LaRoche et al. 1997; McClelland et al. 1997; McClelland and Valiela 1998; Corbett et al. 1999), none of which are in the southeastern United States.

Changes in land use, increased agricultural, industrial, and potable water demands, changing global temperatures, and sea level rise will alter the amount and chemical composition of groundwater entering the coastal zone. In the coastal counties of South Carolina and Georgia, population growth rates are among the highest in the nation (U.S. Census Bureau 2000) and residential and commercial development is escalating. Available data suggest that groundwater inputs to these regions are important (Krest et al. 2000). However, these data are insufficient to evaluate the impact of groundwater-derived fluxes of water, nutrients, and carbon on ecosystem processes in the region. Furthermore, our understanding of how groundwater inputs and groundwater quality will change in response to development is extremely limited. Because of the time required for groundwater in coastal aquifers to travel from areas of recharge to areas of discharge, groundwater quality may integrate anthropogenic inputs over significant temporal (e.g., from decades in confined aquifers to weeks or months in unconfined aquifers) and spatial (e.g., km) scales (Freeze and Cherry 1989). Increased potable water demands on coastal aquifers could lead to withdrawal rates exceeding recharge rates, resulting in saltwater intrusion into shallow aquifers. The expansion of impermeable surface area (e.g., rooftops, highways, parking lots, etc.) will alter runoff regimes, further slowing and/or disrupting natural recharge regimes. Introduction of anthropogenic wastes directly (e.g., septic systems) or indirectly (e.g., infiltration of golf course and agricultural runoff) into surficial aquifers degrades groundwater quality. These changes may significantly alter the role groundwater plays in coastal nutrient budgets.

7.2 Groundwater and Coastal Ecosystems

Ninety-seven percent of the world's freshwater is stored underground as groundwater (Church 1996). Groundwater flows along hydraulic pressure gradients, ultimately leading to points of withdrawal, or more commonly, to surface waters (e.g., streams, lakes, rivers) or to the coastal ocean. Surficial groundwater, in particular, tends to be nutrient rich relative to coastal waters. Despite the importance of the groundwater reservoir, little is known regarding its role as a source of freshwater, nutrients, and dissolved organic material to coastal environments because groundwater inputs to coastal regions were thought to be unimportant. However, recent work (Moore 1996) has corroborated earlier suggestions (Valiela and Teal 1979; Johannes 1980) that groundwater-derived nutrient inputs are important in some coastal systems. For example, groundwater inputs to lakes and rivers have been used to explain high partial pressures of carbon dioxide (CO_2) observed there (Kempe et al. 1991; Mook and Tan 1991). Direct discharge of groundwater into the coastal ocean has been observed (Simmons 1992) and these inputs may exceed river inputs at some sites (Valiela et al. 1978, 1992, 1997; Millham and Howes 1994). Based on dissolved radium isotope (^{226}Ra) data, direct fresh and brackish groundwater discharge to the South Atlantic Bight (SAB) off South Carolina was estimated to be equivalent to 40 percent of the river runoff (Moore 1996). Using dissolved radium budgets in the coastal ocean and dissolved radium, nitrogen (N), and phosphorus (P) concentration relationships in groundwater, Moore et al. (2002) estimated that groundwater-derived nutrient inputs (i.e., loading rates) were potentially as important as, or more important than, river-derived N and P inputs to the South Atlantic Bight.

The potential importance of groundwater-derived nutrient loading to the coastal zone is determined largely by the inputs to the aquifer along the flow path and by the geologic framework of the aquifers through which the groundwater flows. Aquifers of interest to coastal discharge may be separated into two classes, (1) surficial, unconfined aquifers and (2) deep, confined aquifers. Surficial, unconfined aquifers are exposed to recharge from overlying water sources, such as rainfall and irrigation. Recharge occurs throughout the surficial aquifer system and is not limited to recharge areas per se. In coastal regions, water circulates through surficial aquifers relatively rapidly (on the order of weeks to months) and generally flows around marshes but discharges close to shore (Harvey and Odum 1990). In contrast, deeper, confined aquifers are recharged only within recharge zones, which may be located tens to hundreds of miles from discharge zones or water withdrawal points. Water circulates through these confined aquifers much more slowly (on timescales of years to decades) and, depending on the degree of confinement, these aquifers may discharge far from shore (Freeze and Cherry 1989).

In coastal Georgia and South Carolina, the aquifers of interest to coastal discharge are the surficial aquifer and the Upper and Lower Floridan aquifers. All three aquifers are composed of highly permeable deposits separated by impermeable confining units (Bush and Johnston 1988). A detailed study of the surficial aquifer on Sapelo Island (Georgia) indicated that discharge into neighboring estuaries is restricted by marsh deposits. Marshes appear to force groundwater flow through multiple paths; diffuse seepage through marsh sediments, creek bottom baseflow, and deeper submarsh flow (Schultz and Ruppel 2002). The Upper Floridan aquifer is recharged in southwest Georgia, 150 miles from the Atlantic coast. Water flows through deep aquifers over much longer timescales (decades) and may discharge kilometers offshore, as observed for the Castle Hayne aquifer off the coast of North Carolina (Moore et al. 2002). In preindustrial times the hydraulic pressure within the Upper Floridan aquifer was strong enough in Savannah, Georgia, to drive groundwater 10 m above the land surface (Bush and Johnston 1988; Miller 1990), suggesting ample pressure to discharge kilometers offshore.

The region of groundwater discharge into shallow coastal waters, such as estuaries, is often characterized by the mixing of fresh- and salt water in a subterranean estuary (Moore 1999). As a result of this mixing, groundwater that discharges into the coastal zone is not simply fresh groundwater but a combination of groundwater and seawater that is enriched with dissolved materials that derive from the nutrient-enriched groundwater and from reaction of salt water within aquifer sediments (Moore 1999). Even though groundwater discharge may not be an important source of freshwater to the coastal zone per se, it can still be a significant source of dissolved materials to these environments. Determining the magnitude of these material fluxes is an important but difficult task. Groundwater represents an underground, diffuse flow that, like overland flows, connects geographically separate regions and is influenced by a variety of physicochemical parameters. The composition of coastal groundwater represents the net effect of different source terms (e.g., upstream as well as local effects) and internal processing (e.g., the results of microbial processes occurring within the groundwater aquifer). The same processes that affect surface water quality, particularly land use, also impact groundwater quality.

One of the primary differences between surface (e.g., rivers and streams) and groundwater is the relative availability of dissolved oxygen, noted hereafter as O_2. Surface waters are in contact with the O_2-rich atmosphere and atmospheric O_2 freely exchanges via diffusion with these waters. Surface waters also support primary production, mainly via oxygenic photosynthesis, which provides an internal source of O_2 to these waters. Oxygen depletion from surface waters requires high concentrations of labile organic and stratified conditions, which serves to stimulate O_2 consumption (via respiration) and limit O_2 exchange, creating hypoxic or anoxic conditions. Groundwater, on the other hand, is effectively separated from the atmosphere when it enters an aquifer. Isolation of groundwater from the atmosphere means that the groundwater O_2 content is limited to that which is dissolved in the water when it enters the aquifer. Unlike surface waters, groundwater does not support oxygenic photosynthesis because no photosynthetically active radiation (i.e., sunlight) is available, so there is no internal O_2 generation within an aquifer. Additions of O_2 may exist along the aquifer flow path (as oxygenated water enters shallow unconfined aquifers), but microbial processes within the aquifer can rapidly draw down O_2, leading to anoxic conditions (Drever 1982). Dissolved O_2 concentration is therefore a critical biogeochemical variable in subterranean aquifers, as O_2 availability influences the pathway and rates of microbial metabolism and the accumulation of reduced metabolites in the groundwater (Drever 1982).

Microbial processes occurring within an aquifer alter the concentrations and forms of elements along the flow path. The dominant mode of microbially mediated organic matter mineralization (e.g., aerobic respiration vs. denitrification vs. iron reduction vs. methanogenesis) in an aquifer is influenced by the availability and concentration of dissolved O_2, (labile) organic carbon, and the availability of alternate electron acceptors (Drever 1982). The dissolved O_2 concentration reflects the length of time the groundwater has been isolated from the atmosphere, the rate of recharge along the flow path (for shallow, partially confined aquifers), and the rate of O_2 consumption. The availability of labile organic matter (by organic matter, we mean simply organic carbon that may be metabolized by microbes; we are not referring to any particular compound or suite of compounds) also influences microbial metabolic rates (Drever 1982). Dissolved organic matter availability is influenced by surface inputs and possibly by leaching of organic matter from the solid phase as water transits an aquifer. Oxygen concentrations also impact nutrient concentrations and speciation in groundwater. For example, inorganic phosphate is attached to iron oxyhydroxide particles under oxic conditions and dissolved phosphate concentrations may be low. Under anoxic conditions, in contrast, microbially mediated iron reduction occurs, releasing reduced Fe (Fe^{2+}) and inorganic phosphate (Lovley et al. 1990). Similarly, under oxic conditions, ammonia may be converted to nitrate and nitrate accumulates. Under anoxic conditions, nitrate is reduced to dinitrogen (Chapelle 1992).

In the Floridan Aquifer, for example, nitrate reduction appears to be widespread. Concentrations of nitrate and nitrite are higher at recharge areas (50 ± 77 µM in unconfined vs. 18 ± 56 µM confined) than at discharge areas (5 ± 11 µM unconfined, 7 ± 40 µM confined), suggesting consumption, presumably by denitrification, during transit through the aquifer. The trend for ammonium, where data are available, is reversed, with concentrations lower at unconfined recharge areas (3.6 ± 5 µM) than at unconfined discharge areas (28 ± 16 µM; Sprinkle 1989). Ammonium is a by-product of microbial metabolism, so this N species should accumulate during transit through the aquifer. Internal processing occurs during groundwater transit through an aquifer and influences the geochemistry of the groundwater potentially entering the coastal zone.

A variety of physical parameters influence the transport of groundwater to coastal systems (Figure 7.1). Understanding the relationship between groundwater-derived materials and coastal eutrophication requires information regarding the physicochemical, geochemical, and microbiological parameters within and upstream of coastal aquifers. Nutrient (N, P), organic matter, and trace gas concentrations in groundwater are affected by many factors, including land use patterns; water residence time, which is affected by the recharge rate and the hydraulic conductivity of the aquifer system; mixing of different water types; and biogeochemical processes occurring within the aquifer (Freeze and Cherry 1989). The depth to the water table and the degree of confinement are also important, since they influence how rapidly the aquifer is affected by changes in surface inputs: shallow, partially confined, surficial aquifers may be recharged throughout the

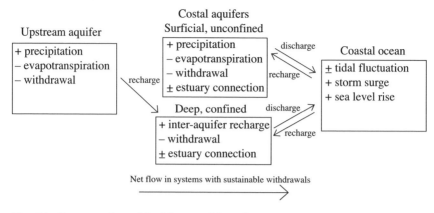

Fig. 7.1. Conceptual model of the variables affecting groundwater flux for a hypothetical system composed of two coastal aquifers with an open connection to an estuary. Discharge is driven by hydraulic gradients but restricted by confining units and impermeable sediments. + indicates increased hydraulic pressure in that component of the model and increased efflux along available flow paths. − indicates decreased hydraulic pressure in that component of the model and decreased efflux along available flow paths. ± indicates variable (either + or −) effects.

flow path, whereas deep, confined aquifers are affected primarily at the upstream recharge site.

Knowing the geochemical signature of groundwater (e.g., oxygen, nutrient and organic matter content) and the groundwater-derived flux of materials to coastal waters permits correlation of groundwater fluxes and observed biological impacts in coastal receiving waters. The complexity of such linkages is illustrated in a data set describing the variability of groundwater inputs and the occurrence of a brown tide in Long Island Sound (LaRoche et al. 1997). Variations in groundwater discharge were correlated with changes in the ratio of dissolved organic nitrogen (DON) to dissolved inorganic nitrogen (DIN). Groundwater was enriched in DIN relative to DON, and increasing groundwater inputs lowered the DON:DIN ratio in sound waters. Variability in the DON:DIN ratio influenced the timing and magnitude of blooms of the brown tide alga, *Aureococcus anophagefferens*. Blooms of *A. anophagefferens* were observed in coastal bays characterized by a higher ratio of DON:DIN. In years with high groundwater discharge, the DON:DIN ratio in the bays was low, and the growth of *A. anophagefferens* was overshadowed by phytoplankton that preferred DIN. In dry years, groundwater discharge and associated DIN inputs were lower and at these times, DON dominated the total nitrogen pool, leading to *A. anophagefferens* blooms.

In order to evaluate the role of groundwater in coastal eutrophication, one must address several fundamental questions. First and foremost, a baseline of groundwater flux and concentration data must be established against which future change can be gauged. At present, baseline data are sparse, particularly in Georgia. Such data are required to make sound management decisions in the future. Second, the sources of nutrients and organic materials in groundwater must be determined to quantify the link between groundwater biogeochemistry and land use. Third, the importance of dissolved organic nutrients, DON and dissolved organic phosphorus (DOP), in groundwater should be determined and their bioavailability to primary and secondary producers in coastal ecosystems must be assessed. Fourth, the ability of aquifer microbes to alter concentrations and forms (i.e., bioavailability) of organic materials *before* they are delivered to coastal environments must be evaluated. Fifth, data on source and availability must be merged with estimates of groundwater flux to assess the contribution of groundwater to coastal nutrient and materials budgets. Finally, a dynamic mechanistic model relating groundwater nutrient loading to coastal land use must be developed. These questions are addressed in more detail later in this chapter.

7.3 Available Data Describing Carbon, Nitrogen, and Phosphorus Concentrations in South Carolina and Georgia Groundwater

The data presented below have been compiled from the peer-reviewed literature, federal and state government reports, and unpublished sources. The depth, or depth range, that was sampled is noted because the depth of groundwater sampled varied between studies. The magnitude of groundwater-derived inputs, relative to river-derived inputs, to coastal systems varies substantially (Table 7.1). However, groundwater is a globally significant input that averages about 20 percent of riverine input on a worldwide basis (Zekster and Loaiciga 1993) and up to 50 percent of riverine input in some closely studied systems (Valiela et al. 1992). In general, significant amounts of dissolved and particulate material are transported via groundwater because the concentrations

Table 7.1. Estimates of Groundwater Input to the Coastal Ocean.

Location	% of riverine input	Reference
Chesapeake Bay	10	Hussain et al. 1999
South Carolina	40	Moore 1996
Western Australia	20	Johannes 1980
Waquoit Bay	50	Valiela et al. 1992
Worldwide	20	Zekster and Loaiciga 1993

of solutes are usually much higher in salty groundwater than in river or coastal waters (Valiela et al. 1978, 1990; Moore 1999). Therefore, the biological (e.g., increased rates of primary production or bacterial respiration) and ecosystem-level (e.g., change in species distribution of phytoplankton, increased frequency of harmful algal blooms, or water column hypoxia/anoxia) repercussions of groundwater-mediated chemical inputs to coastal ecosystems can be significant (Church 1996). Information pertaining to the distribution of carbon (C), nitrogen (N), and phosphorus (P) species in groundwater is presented below. For N and P, data describing dissolved inorganic (nitrate, NO_3^-, nitrite, NO_2^-, ammonium, NH_4^+, ortho-phosphate, HPO_4^{2-}) and organic (DON and DOP) forms as well as gaseous N species (nitrous oxide, N_2O) are discussed. For carbon, data describing dissolved inorganic (DIC) and organic (DOC) C, and gaseous C (methane, CH_4) are presented.

7.3.1 Dissolved Inorganic and Organic Nitrogen

Groundwater transport of N to coastal aquifers is far more significant and widespread than previously thought, primarily due to anthropogenic alterations within coastal watersheds (Valiela et al. 1992, 1997). Groundwater flows into coastal systems could provide a quantitatively significant source of N, which may impact eutrophication because primary production in many coastal ecosystems is N limited (Ryther and Dunston 1971; Legendre and Gosselin 1989; Nixon 1995; Paerl 1997; Valiela et al. 1997). Changes in groundwater-derived inorganic N fluxes could, therefore, increase primary production and contribute to coastal eutrophication. Coastal waters in Georgia and South Carolina often have elevated NO_3^- concentrations, up to 35.7 μM. However, few data are available for shallow groundwater in coastal South Carolina and Georgia (Table 7.2). In other coastal regions, NO_3^-, or $NO_3^- + NO_2^-$,

Table 7.2. Nitrate Concentrations in Coastal Aquifers.

Location	Range (μM)	Sampling Depth (feet)	Source
Bly Creek Basin, SC	1.93	Surficial	Dame et al. 1991
Cape Cod, MA	0–450	not reported	Frimpter and Gay 1979
Coastal Plain, GA/FL	0–543	Surface to 1000	Sprinkle 1989
Coastal Plain, GA/SC	<0.7–130	10–200	Burt 1993
Coastal Plain, NC	<0.7–2250	surficial	Gilliam et al. 1974
Coastal Plain GA (n = 41)	0.7–1000	surface to 1500	Earthinfo 1997
Coastal SC (n = 381)	0.7–450	surface to 1500	Earthinfo 1997
Delmarva Peninsula, VA	6.4–650	Surficial	Reay and Gallagher 1992
Falmouth, MA	0.7–694	not reported	Meade and Vaccaro 1971
Florida Keys	<0.7–2.8	domestic wells	LaPointe et al. 1990
Long Island, NY	7.86	not reported	Bowman 1977
Murrels Inlet, SC	0–6.42	Surficial	Aelion et al. 1997
North Inlet, SC	5.71	Surficial	Aelion et al. 1997
Wye River, MD	957–1164	Surficial	Staver and Brinsfield 1996
Orleans, MA	0–392	not reported	Teal et al. 1982
Tomales Bay, CA	<0.7–2.85	domestic wells	Oberdorfer et al. 1990

concentrations can be even higher (up to 2,250 μM; Gilliam et al. 1974). In highly populated areas where septic tanks are the primary form of sewage treatment, inorganic N rich materials enter the groundwater through leach fields, and groundwater NO_3^- concentrations as high as 714 μM have been documented (Table 2; Meade and Vacarro 1971).

Direct measurements of groundwater discharge into coastal environments indicates that the NO_3^- content of groundwater discharging to coastal waters may be orders of magnitude greater than the concentration in the receiving coastal waters (Table 7.3). For example, NO_3^- concentrations in offshore coastal waters tend to be low, ranging between 0.07 to 0.5 µM (Paerl 1997). In contrast, NO_3^- concentrations in estuarine waters range from 0 to 100 µM, and elevated concentrations in the upper reaches of estuarine systems could reflect substantial groundwater inputs (Fisher et al. 1988). Most of this DIN is taken up rapidly within the estuary, and very little dissolved inorganic N is actually delivered to the coastal zone (Fisher et al. 1988). While a clear connection between population density, septic tanks, and NO_3^- contamination of groundwater has been shown in some regions (Valiela et al. 1992), correlations between groundwater NO_3^- concentrations and residential land use in South Carolina and Georgia cannot be made at this time because systematic data delineating residential land use types is insufficient (see below). However, the impact of agriculture on groundwater nitrogen concentrations in the Georgia coastal plain has been examined. In southwest Georgia, agricultural activities increased NO_3^- concentrations in surficial groundwater to a median value of

Table 7.3. Nitrate concentrations in groundwater discharging into coastal systems.

Location	Range (µM)	Reference
Agana Bay, Guam	178.6	Marsh 1977
Discovery Bay, Jamaica	85.7–250	D'Elia et al. 1981
Great Sippewissett Marsh, MA	10–100	Valiela et al. 1978
Swan River Estuary, W Australia	114.3–378.6	Johannes 1980
Town Cove, MA	7.1–107.1	Giblin 1983
Town Cove, MA	57.1–200	Giblin and Gaines 1990
Tumon Bay, Guam	71.4–142.9	Matson 1993
Western HI Island	28.6–92.8	Kay et al. 1977

479 μM (Crandall 1996). Increased concentrations of NO_3^- in surficial aquifers have been linked directly to an increase in NO_3^- in Suwanee River waters, also in Southwest Georgia (Pittman et al. 1997).

Ammonium (NH_4^+) is another inorganic form of N found in groundwater, although NH_4^+ concentration data are less frequently reported than NO_3^- (Table 7.4). In particular, NH_4^+ can be a dominant form of DIN in anaerobic or suboxic aquifers. For example, in North Inlet, South Carolina, NH_4^+ concentrations can be as high as 250 μM as NH_4^+-N (Table 7.4). Generally speaking, however, NH_4^+ concentrations are lower than NO_3^- concentrations. This is likely a function of the O_2 concentration of the aquifer. Nitrate would be expected to be the dominant N form in aerobic groundwater, while NH_4^+ would be expected to be the dominant N form in anaerobic ground water. Data with respect to land use and NH_4^+ concentrations are sparse. However, available data from the Delmarva Peninsula, Virginia, show that groundwater NH_4^+ concentrations in agriculture-dominated and forest-dominated regions were similar (Table 7.4; Reay and Gallagher 1992).

While NO_3^- and NH_4^+ are the N forms most often considered when discussing groundwater N sources, dissolved organic nitrogen, DON, is another potentially important N pool in groundwater. DON is the major dissolved N species in rivers in the Georgia-Florida coastal plain (Ham 1997) and estuarine waters in Georgia typically have comparable concentrations of DON and DIN (Joye unpublished data). Traditionally, the DON pool has been viewed as a large refractory pool that persists in marine systems, which are often N limited, because its rate of utilization is very low. However, over the last ten years, the DON pool has been shown to be more labile than previously thought (reviewed in Bronk 2002). For example, Bronk and Glibert (1993) found that uptake rates of DON rivaled or even exceeded DIN uptake rates in the Chesapeake Bay. Seitzinger and Sanders (1997) performed bioassays with DON from a variety of natural and anthropogenic sources and found that ~70 percent of the DON pool was utilized on timescales of 14 days or less. Though data are scarce, DON concentrations as high as 93 μM in coastal groundwater have been observed (Table 7.5). DON concentrations in the Upper Floridan aquifer (median 14.3 μM), which is in open connection with surface waters in much of Florida, are similar to DIN concentrations (median 20.7 μM, Sprinkle 1989). Given that a significant fraction of the DON pool in surface waters appears to be bioavailable on timescales of days, groundwater DON is likely an important and underestimated source of bioavailable N to coastal plankton.

Table 7.4. Ammonium Concentrations in Coastal Aquifers.

Location	Range (μM)	Sampling Depth (feet)	Source
Coastal GA (n=5)	0.7–5	Surface to 1500	Earthinfo 1997
Coastal SC (n=319)	0.07–185.7	Surface to 1500	Earthinfo 1997
Coastal Plain, GA/FL	0–150	Surface to 1000	Sprinkle 1989
Dog Creek, SC[1]	7.14–50	Surficial	Aelion et al. 1997
Oyster Creek, SC[2]	0–178.6	Surficial	Aelion et al. 1997
Coastal Plain, GA/SC	0.7–192.8	10-200	Burt 1993
North Inlet, SC	28.6–250	Surficial	Gardner pers. Com.
Bly Creek Basin, SC	35.7	Surficial	Dame et al. 1991
Delmarva Peninsula, VA[3]	2.8–9.3	Surficial	Reay and Gallagher 1992
Delmarva Peninsula, VA[2]	5.7–12.8	Surficial	Reay and Gallagher 1992
Tomales Bay, CA	0.7–714	Domestic Wells	Orberdorfer et al. 1990
Cape Cod, MA	0–64.3	Springs	Valiela et al. 1990
Great Sippewissett Marsh, MA	2.1–12.1	Springs	Valiela et al. 1990
Florida Keys	0.7–2.86	Surficial	LaPointe et al. 1990

[1] suburban

[2] forested

[3] agriculture

7.3.2 Dissolved Inorganic and Organic Carbon

Dissolved inorganic C (DIC) is another important constituent that is carried by groundwater to the coastal zone. Data on DIC concentrations in groundwater are scarce (Table 7.6). Reported groundwater

Table 7.5. Dissolved Organic Nitrogen Concentrations in Groundwater.

Location	Range (μM)	Depth (feet)	Source
Coastal GA (n=3)	10–28.6	To 1500	Earthinfo 1997
Coastal Plain, GA/FL	0–86	Surface to 1000	Sprinkle 1989
Bly Creek Basin, SC	92.8[1]	Surficial	Dame et al. 1991
Great Sippewissett Marsh, MA	35.7–78.6[2]	Surficial	Valiela et al. 1978
Childs River, MA	42.8 ± 14.3[1]	Top of water table	Rudy et al. 1994
Sage Lot Pond, MA	107 ± 14.3[1]	Top of water table	Rudy et al. 1994

[1] persulfate oxidation
[2] Kjeldahl (TON)

DIC concentrations for Georgia and South Carolina coastal regions range from 0 to 12 mM. These data represent compilations for a variety of depths (to 500 m) and are regionally scattered. In one of the few studies specifically investigating DIC in groundwater, Cai et al. (2003) observed North Inlet (South Carolina) groundwater had much higher DIC concentrations than adjacent seawater or river water. DIC values in groundwater, river water, and seawater were 8 to 12 mM, 0.4 to 0.8 mM, and 2.1 mM, respectively. Although groundwater DOC data span a similar concentration range, 0 to 4.1 mM (Table 7.7), whether it is source of potentially labile DOC to coastal systems has not been determined. Labile dissolved organic matter inputs could stimulate heterotrophy in coastal ecosystems and exacerbate hypoxic or anoxic conditions in coastal regions.

7.3.3 Dissolved Inorganic and Organic Phosphorus

Similar to N, P availability may be an important control on primary production in coastal ecosystems, and, at times, P availability may limit production in the coastal zone (Malone et al. 1996; Paerl 1997). The interplay between N and P as limiting or colimiting nutrients in coastal systems is dynamic. Dissolved inorganic P (present as HPO_4^{2-} at

Table 7.6. Dissolved Inorganic Carbon (as HCO_3^-) Concentrations in Groundwater.

Location	Range (mM)	Depth (feet)	Source
Coastal GA (n=52)	0.17–4.7	To 1500	Earthinfo 1997
Coastal Plain SC	11	To 1500	Earthinfo 1997
Coastal Plain GA/SC	1.8–12.5	10–200	Burt 1993
Black Creek aquifer, SC	0.1–12	Deep aquifer	Chapelle and McMahon 1991
Tomales Bay, CA	0.58–6.2	Domestic well	Orberdorfer et al. 1990
North Inlet, SC	2–12	Surface	Cai et al. 2003
River Water	0.4–0.8	Surface	Cai et al. 2003
Seawater	2.8	Surface	Cai et al. 2003

Table 7.7. Dissolved Organic Carbon (DOC) Concentrations in Groundwater.

Location	Range (μM)	Depth (feet)	Source
Coastal Plain GA/SC	16.6–433	10–200	Burt 1993
Florida and Georgia	8.3–417	106–1000	Leenheer et al. 1974
Coastal Plain, GA/FL	8–3330	Surface to 1000	Sprinkle 1989
Coastal Georgia (n = 4)	16.6–75	To 1500	Earthinfo 1997
Coastal SC (n = 284)	33–4083	To 1500	Earthinfo 1997
Savannah River, SC	25–167	Surficial	Dosskey and Bertsch 1997
Bly Creek Basin, SC	767	Surficial	Dame et al. 1991
Southern Ontario, CA	75–1500	141–525	Aravena and Wassenaar 1993

Table 7.8. Dissolved Phosphorus Concentrations in Groundwater.

Location	Range (µM)	Sampling Depth (feet)	Source
North Inlet SC	0.1–77.4[1]	Surficial	Gardner pers. comm.
Coastal Georgia (n = 4)	0.3–0.6[1]	Surface to 1500	Earthinfo 1997
Coastal SC	161[1]	Surface to 1500	Earthinfo 1997
Coastal Plain, GA/FL	0–39[1]	Surface to 1000	Sprinkle 1989
Coastal Plain, GA/FL	0–58[2]	Surface to 1000	Sprinkle 1989
Bly Creek Basin, SC[3]	2.9[1]	Surficial	Dame et al. 1991
Bly Creek Basin, SC[3]	4.2[2]	Surficial	Dame et al. 1991
Delmarva Peninsula, VA[4]	0.3–0.58[1]	Surficial	Reay and Gallagher 1992
Delmarva Peninsula, VA[5]	0.19[1]	Surficial	Reay and Gallagher 1992
Tomales Bay, CA	0.09–54.8[1]	Domestic Wells	Orberdorfer et al. 1990
Cape Cod, MA	0.3–16.1[1]	Springs	Valiela et al. 1990
Great Sippewissett Marsh, MA	0–10[1]	Springs	Valiela et al. 1990
Florida Keys	<0.03–106[1]	Surficial	LaPointe et al. 1990

[1] DIP: dissolved inorganic orthophosphate
[2] TP: total phosphorus
[3] annual average
[4] agriculture
[5] forested

the pH of most groundwater) concentrations in groundwater vary over three orders of magnitude (0.32 to 161 µM; Table 7.8). As in the case of NO_3^- versus NH_4^+, concentrations of HPO_4^{2-} in groundwater are related to the redox regime. Under oxic conditions, HPO_4^{2-} may be partitioned onto the solid phase via sorption to reactive iron oxyhydroxides, limiting the potential flux of HPO_4^{2-} through aquifers with high adsorption capacities. However sediments, predominantly sands, that have a low adsorption capacity are common on the Atlantic coastal plain (National Research Council 2000). Under anoxic conditions, iron oxyhydroxides are reductively dissolved, thereby releasing HPO_4^{2-} into solution. High concentrations of dissolved HPO_4^{2-} would therefore be expected in hypoxic or anoxic groundwater, while low concentrations of dissolved HPO_4^{2-} would be expected in oxic groundwater. Unfortunately, groundwater O_2 concentrations were not presented in most of the papers cited in Table 7.8. Furthermore, most of the P data was collected from random depths. More data on groundwater P concentrations and loading rates are needed to evaluate the potential importance of groundwater-derived P in coastal nutrient dynamics.

Dissolved organic P (DOP) is often available for uptake by microorganisms (Cembella et al. 1984; Rivkin and Swift 1985; Bjorkman and Karl 1994). When both dissolved inorganic P (DIP) and DOP pools are quantified, DOP concentrations frequently exceed DIP (Butler et al. 1979; Jackson and Williams 1985; Karl and Tien 1997). DOP is rapidly turned over enzymatically and taken up as DIP in the pelagic environments (Rivkin and Swift 1985; Orrett and Karl 1987). Therefore, groundwater-derived DOP could be bioavailable and serve as a significant P source to coastal primary producers. Unfortunately, data describing DOP concentrations in groundwater are sorely lacking; we were unable to identify any references on this topic.

7.3.4 Trace Gases: N_2O and CH_4

In addition to inorganic and organic nutrients, groundwater may also contain high concentrations of dissolved gases. Of these dissolved gases, some are common bimolecular gases (e.g., O_2, N_2) while others are radiative trace gases (e.g., N_2O and CH_4) that result from microbial metabolism and may contribute to the radiative heating of the atmosphere upon efflux from the aquifer. Nitrous oxide (N_2O) has a long atmospheric residence time (~150 yr) and the destruction of N_2O in the stratosphere contributes to ozone degradation. Nitrous oxide is 300 times more efficient than CO_2 in capturing radiative heat and is thus a more effective greenhouse gas. Methane has a shorter atmospheric residence time (~10 yr) but is 15 times more efficient than CO_2 in capturing radiative heat. Thus, although the atmospheric concentration of N_2O (~300 ppb) and CH_4 (~1.7 ppm) are much lower than that of CO_2 (~350 ppm), both gases contribute significantly to global warming because of their enhanced radiative heating capacity.

Because groundwater is isolated from the atmosphere, O_2 consumption via bacterial respiration in groundwater reduces the O_2 concentration and increases the concentration of metabolic end products. Under low O_2 conditions, the gaseous end products of nitrate respiration (denitrification), such as N_2O and N_2, accumulate in groundwater. Under anoxic conditions, hydrogen sulfide (H_2S) and CH_4 also accumulate (Bugna et al. 1996). When groundwater is discharged to the coastal zone, a significant fraction of the dissolved gas load may be released to the atmosphere. Though N_2O and CH_4 concentrations are often elevated in groundwater (Ronen et al. 1988), the fate of these trace gases in coastal systems is unknown (LaMontagne et al. 2003).

Nitrous oxide is a by-product of both nitrification and denitrification, two microbially mediated processes that transform fixed N between oxidized, reduced, and gaseous forms. No N_2O concentration data were found for coastal (or inland) Georgia or South Carolina groundwater (Table 7.9). However, elevated N_2O concentrations in groundwater have been observed elsewhere. In agriculturally impacted regions, N_2O concentrations were extremely high (approaching 11.4 μM; Table 7.9). Methane concentrations in groundwater may be supersaturated (Table 7.10). Data were also not available for CH_4 concentration in the shallow coastal groundwater of Georgia and South Carolina. As a result, data from other systems are presented. In the deeper aquifers of Georgia (Black Warrior, Chattahoochee River, and Pearl River), CH_4 concentrations are low (< 0.2 μM; Table 7.10). In shallow and deep aquifers around the Gulf of Mexico and throughout Canada, concentrations of CH_4 from 0.06 μM to 4750 μM have been reported. Methanogenesis is an anaerobic process; therefore, high CH_4 concentrations in groundwater imply negligible concentrations of dissolved O_2. The importance of N_2O and CH_4 as radiative trace gases makes determining their source terms important. Given the high concentrations (relative to the atmospheric mixing ratios) of N_2O and CH_4 in groundwater, the groundwater flux term could be globally significant.

7.3.5 Land Use

Loading of inorganic and organic materials to either to ground or surface waters is closely linked to watershed land use patterns (Valiela et al. 1990, 1992). A variety of anthropogenic alterations, including

Table 7.9. Nitrous Oxide Concentrations in Groundwater.

Location	Source	Range (nM)	Depth (ft)	Reference
Hubbard Brook, NH	pristine forest	5.7–13.6	Springs	Bowden and Bormann 1986
Hubbard Brook, NH	recent clear cut forest	9–11273	Springs	Bowden and Bormann 1986
Central Platte Valley, ID		0.2–6182	to 30	Spalding and Parrott 1994
Cambridgeshire, England	Agricultural	54.5–1227	Surficial	Mühlherr and Hiscock 1997
Western Tokyo	Septic waste impacted	36.4–454	Springs	Ueda et al. 1991
Western Tokyo		13.6–318	450–750	Ueda et al. 1991
S. England	Agricultural drains	22.7–3000	Surficial	Dowdell et al. 1979
Veluwe, Netherlands	Acid rain and manure	250–682	27	Ronen et al. 1988
Jerusalem, Israel	Municipal effluent	91–295	Surficial	Ronen et al. 1988
Coastal Aquifer, Israel	Effluent used to irrigate	1931–9091	450–750	Ronen et al. 1988
Waquoit Bay, MA	Septic	0.7–3182		LaMontagne et al. 2003
Coastal Sea and Ocean		trace to 14		Ronen et al. 1988
Unpolluted Estuary		9.1–45		Ronen et al. 1988

Table 7.10. Methane Concentrations (μM Unless Otherwise Noted) in Groundwater.

Location	Range	Sampling Depth (feet)	Source
Black Warrier River Aquifer	0–18.8[1]	>1500 (coast)	Lee 1993
Chattahoochee River Aquifer	0–5[1]	>1500 (coast)	Lee 1993
Pearl River Aquifer	0–3.8[1]	>1500 (coast)	Lee 1993
Gulf of Mexico	0–203	12–171m	Bugna et al. 1996
Canada/US	0–4725	<300	Barker and Fritz 1981
Texas Cretaceous aquifer	0.06–1219		Zhang et al. 1997
Texas Eocene aquifer	0.06–3.1		Zhang et al. 1997
Southern Ontario	7.5–3512	50–425	Valiela et al. 1990
Eastern Canada	0.06–330	<300	LaPointe et al. 1990

[1] nM

urbanization, agricultural practices (e.g., pastureland and row crops), and recreational development (e.g., golf courses), contribute to alterations in both the amount and composition (nutrient and organic loads) of freshwater delivered to coastal ecosystems. The influence of land use on river and lake water quality is easily observed because these waters are visible and in close contact to both scientists and the general public. The influence of anthropogenic alterations on groundwater is not clear, and studies of groundwater-mediated transport have been less common because these waters are underground and less obviously impacted.

River flow is controlled largely by precipitation, but it is also affected by development along the river, e.g., deforestation, construction of dams, and municipal and industrial diversion. Deforestation alters the sediment load and the forms of nutrients delivered to rivers from the watershed (Valiela et al. 1992). Construction of dams along rivers alters the sediment load as well as the distribution of sediment types (e.g., fine vs. coarse grained materials; Humborg et al. 1997). Water diverted for municipal and industrial use typically ends up back in the river, but in a highly altered form (in particular, nutrients loads may be elevated). Groundwater is impacted by anthropogenic forces in much the same way. For example, Valiela et al. (1992) established a linear correlation between watershed development (number of buildings per hectare) and groundwater NO_3^- concentration (see Figure 3 of Valiela et al. 1992), suggesting that groundwater nutrient loads increase proportionately to urbanization. Agricultural impacts result in similar correlations. Though some studies have examined the correlation between agricultural (e.g., row crops and animal industry) activity and surface and ground water nutrient concentrations (Girard and Hillaire-Marcel 1997; Iqbal et al. 1997; Jordan et al. 1997a, 1997b; Mallin et al. 1997; Matson et al. 1997), detailed studies of agricultural impacts on groundwater are lacking. Such studies are sorely needed as proposals for expansion of agricultural activity, particularly intensive animal operations, in coastal Georgia are pending. There are currently no baseline data against which to gauge the magnitude of future alterations of groundwater nutrient concentrations that could result from the proposed industry expansion.

In a simplified sense, estuaries, either directly or indirectly (through marshes), collect chemicals released from freshwater inputs, which are derived from both surface and groundwater flows. Estuarine residence time can influence whether chemicals derived from the land are transported to the coastal ocean or processed by biological activity within the estuary. For example, Cai et al. (2000) found that the seasonal difference in the abundance of nitrate and nitrite in the Satilla River, Georgia, could be explained by changes in river flow, assuming marsh respiration rates are constant. Concentrations of $NO_3^- + NO_2^-$ were higher during the low-flow summer period, when the water residence time is long, than during the high-flow period of April, when the residence time is short. This is also true of P in Georgia estuaries (Pomeroy et al. 1972). Impoundment of rivers and coastal barrier islands and regulation (diversion) of river water can also affect the balance between surface and groundwater inflows. Fortunately, both Georgia and South Carolina salt marshes have been protected from impoundment and infilling since the 1970s (Smith 1988). Despite the passage of the Coastal Marshland Protection Act in 1970, more than 4,330 hectares of marshland have been destroyed nationally as a result of road construction and dredge and fill activities (Kundell et al. 1988).

Sources of anthropogenic inorganic nutrients to coastal aquifers include agricultural fertilizers, animal wastes, golf courses, septic tanks, and sewage treatment facilities, as well as airborne pollutants (e.g., NH_4^+) added to groundwater via rainfall and recharge. With respect to agriculture, fertilizer use has increased substantially. In 1938, Georgia farms used approximately 696,717 metric tons of N fertilizer; by 1994, N fertilizer usage had risen to 1,379,827 metric tons (Irvin et al. 1994). The growth of golf courses along the South Carolina and Georgia coast and on barrier islands represents another mechanism of increasing the flux of inorganic N and P because of the heavy fertilizer use (Fig. 7.2); newer commercial fertilizer mixtures contain high concentrations of urea, an organic N form not monitored routinely. With an average fertilization rate of 75 kg per acre, the impact of golf course runoff on groundwater nutrient concentrations could be substantial. Groundwater N and P concentrations around golf courses are elevated.

Data describing land use type and NO_3^- and HPO_4^{2-} concentrations in Georgia groundwater are presented in Table 7.11 (data provided by Prof. P. Bush of the University of Georgia). The data were collected using a nonrandom experimental design where individuals interested in knowing the concentrations of nutrients in their well water contributed samples for analysis. As such, the data were obtained from drinking water wells for the most part and are not necessarily representative of the groundwater impacted by various animal industries. A final consideration is that the coastal plain is poorly represented in the data set. Most of these data are from the interior, central portions of the state, and the depths of the wells sampled varied substantially. No significant concentration-depth correlations were observed within the data, so the data were combined and sorted according to land use. Wells characterized as being impacted by households consistently exhibited the highest concentrations of NO_3^- and HPO_4^{2-} (Table 7.11). Poultry and swine impacted regions exhibited higher NO_3^- concentrations. These data underscore the need for a more systematic survey of coastal groundwater in order to document correlations between land use and groundwater nutrient loads. Such studies should aim for a consistent sampling regime (e.g., depth sampled) and a random sampling grid (including the aquifers downstream from settling ponds, as

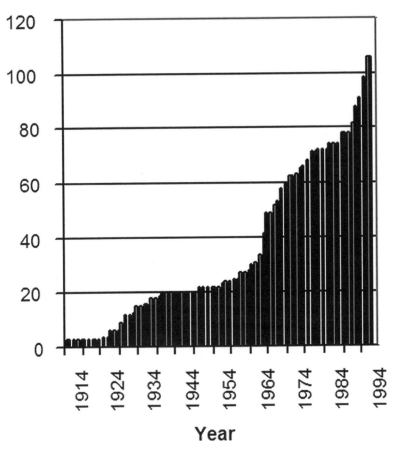

Fig. 7.2. Trends in total number of golf courses in the state of Georgia that lie within drainage basins with access to the coastal zone. Data were normalized for courses with nine holes. Compiled from data in McCollister (1993).

well as aquifers undergoing sea water intrusion, etc.). To add to the difficulty of understanding groundwater inputs to coastal zones, even where nutrient data are available and groundwater flow paths are well studied, predictions of coastal nutrient loading via groundwater by different models are highly variable (Valiela et al. 2002).

Table 7.11. Nitrate and phosphate concentrations (µM) in groundwater as a function of land use.

Location	Form	Source	Range	Median	n	Date	Reference
GA well water	NO_3^-	Beef/Dairy	1.4–3214.3	185.7	213	1994–96	Bush pers. comm.
GA well water	PO_4^{3-}	Beef/Dairy	1.9–12.9	3.2	213	1994–96	Bush pers. comm.
GA well water	NO_3^-	Poultry	14.3–5516	342	540	1994–96	Bush pers. comm.
GA well water	PO_4^{3-}	Poultry	3.2–219	3.2	540	1994–96	Bush pers. comm.
GA well water	NO_3^-	Swine	14.3–5500	14.3	84	1994–96	Bush pers. comm.
GA well water	PO_4^{3-}	Swine	3.2–12.3	3.2	84	1994–96	Bush pers. comm.
GA well water	NO_3^-	household	0–17143	171.4	3300	1994–96	Bush pers. comm.
GA well water	PO_4^{3-}	household	0–7620	1.9	10000	1994–96	Bush pers. comm.
W central FL	PO_4^{3-}	Citrus	0.3–74.3	–	–	1990–91	Stanley et al. 1995
W central FL	PO_4^{3-}	Native	1.9–7620	–	–	1990–91	Stanley et al. 1995
Coastal VA	NO_3^-	Agriculture	222–652	–	3	1991	Reay and Gallagher 1992
Coastal VA	NO_3^-	Forested	6.4–78	–	2	1991	Reay and Gallagher 1992
Hawaii	NO_3^-	Golf course	57–94	77	10	1988–90	Dollar and Atkinson 1992
Hawaii	PO_4^{3-}	Golf course	1–4.2	3	10	1988–90	Dollar and Atkinson 1992
W. Australia	NO_3^-	Urban	71.4–3928.6	–	60	1979	Sewell 1982
W. Australia	PO_4^{3-}	Urban	0–12.9	–	67	1979	Sewell 1982

7.4 Recommendations

To evaluate the role of groundwater as a nutrient source in the southeastern United States, an integrative research strategy is needed. Six research areas are highlighted in order from more- basic to more-sophisticated data requirements (1 to 4). These are logical first steps towards understanding the magnitude and variability of groundwater-derived nutrient inputs. The final two research areas are interdisciplinary objectives and are only possible once monitoring data become available.

1. *Establish a baseline of concentration and flux data against which future change can be gauged.* Baseline data describing nutrient, organics, and trace gas concentrations in coastal South Carolina and Georgia groundwater are extremely limited, and detailed seasonal studies throughout the coastal zone are sorely needed. Such data are critical for making sound management decisions.

2. *Determine the sources of nutrients and organic materials in order to quantify linkages between groundwater concentrations and land use.* Linking specific land use(s) with alterations of groundwater nutrients, organics, and trace gas concentrations and groundwater fluxes to coastal systems can be accomplished using modeling and geochemical methods. Linkages between land use and groundwater alterations can only be made once baseline data are available. After establishing the baseline, geochemical methods, such as stable isotope tracers, can be used to trace sources from the land to groundwater aquifers (e.g., Girard and Hillaire-Marcel 1997) or from groundwater aquifers into coastal estuaries (e.g., McClelland and Valiela 1998).

3. *Assess the bioavailability of groundwater DOC, DON and DOP to evaluate the impact on coastal food webs.* Inorganic nutrients are assimilated rapidly by primary producers and bacteria. Organic nutrient sources may also be bioavailable. Determining the magnitude of the groundwater-derived organic N and P source term is important, but assessing the bioavailability of this pool is imperative when trying to estimate the potential impacts of eutrophication. Very few data on groundwater dissolved organic pools exist, and at present, the importance of these potentially significant nutrient sources cannot be estimated. Evaluating the lability of the DOC pool is also important because DOC may stimulate respiration and contribute to coastal hypoxia or anoxia.

4. *Evaluate the ability of aquifer microbes to alter concentrations and forms (i.e. its bioavailability) of materials before they are delivered to coastal environments.* Given that coastal groundwater travels at slow velocities (about 0.3 m d^{-1} in a sandy aquifer, though velocities vary substantially among and between sites; Valiela et al. 1992), there is significant opportunity for microbial alteration as groundwater transits through the aquifer. Over the short term (days-months), aquifer microbes could significantly alter the concentrations and forms of materials present in groundwater. The activity of microbes could reduce the loads of certain components (e.g., NO_3^-) while increasing the loads of others (N_2O). Understanding the role groundwater microbes play in altering groundwater nutrient, organics, and trace gas concentrations in coastal, and possibly in upstream, aquifers is therefore vital.

5. *Merge source and availability data with hydrological estimates of groundwater flux to assess the contribution of groundwater to coastal eutrophication.* To obtain a quantitative estimate of groundwater-derived fluxes to coastal environments, land use, hydrologic, geochemical, and ecological data must be merged. Combining such data will permit the calculation of groundwater-specific loading rates. These data can be compared with riverine and other (atmospheric, oceanic) loading rates in order to evaluate the contribution of groundwater to coastal eutrophication.

6. *Develop a mechanistic model relating groundwater nutrient loading to land use in coastal environments.* Ultimately, the data obtained from addressing Recommendations 1 through 5 could be input into a mechanistic model that would eventually allow predictions to be made of how changes in land use (e.g., urbanization) will change the inputs of materials to coastal systems. Then, the manner in which these inputs might be manifest in altered regional patterns of primary production and nutrient cycling could be assessed.

7.5 Conclusions

Concern over nutrient (primarily N) and organic matter loading to coastal watersheds is increasing because loading is increasing and rates of primary production in coastal waters are largely limited by N supply (Howarth 1988; Valiela 1995). Eutrophication of coastal waters, often linked to increasing N loading from the adjacent watersheds, is arguably the principal and most pervasive anthropogenic alteration to coastal ecosystems across the globe (GESAMP 1990). The contribution of groundwater-derived inorganic nutrients and organic materials to coastal eutrophication is largely unknown but potentially important (Valiela et al. 1992). Determining whether groundwater plays a significant role in the eutrophication of coastal systems is, therefore, a necessary endeavor. In order to properly manage water resources, it is imperative that we have in place a firm understanding of the role of groundwater in the budgets of C, N, and P in coastal environments.

Acknowledgments

Financial support for preparation of this paper was supported by grants from the South Carolina Sea Grant Consortium, the Land Use-Coastal Ecosystems Study (under NOAA Coastal Ocean Program grant NA960 PO 113), and by the Georgia Sea Grant Program.

References

Aelion, C. M., J. N. Shaw, and M. Wahl. 1997. Impact of suburbanization on ground water quality and denitrification in coastal aquifer sediments. *Journal of Experimental Marine Biology and Ecology* 213:31–51.

Aravena, R. and L. I. Wassenaar. 1993. Dissolved organic carbon and methane in a regional confined aquifer, southern Ontario, Canada: Carbon isotope evidence for associated subsurface sources. *Applied Geochemistry* 8:483–493.

Barker, J. F. and P. Fritz. 1981. The occurrence and origin of methane in some groundwater flow systems. *Canadian Journal of Earth Science* 18:1802–1816.

Beaulac, M. N. and K. H. Reckhow. 1982. An examination of land use-nutrient export relationships. *Water Resources Bulletin* 18:1013–1022.

Bjorkman, K. and D. M. Karl. 1994. Bioavailability of inorganic and organic phosphorous-compounds to natural assemblages of microorganisms in Hawaiian coastal waters. *Marine Ecology Progress Series* 111:265–273.

Bowden, W. B. and F. H. Bormann. 1986. Transport and loss of nitrous oxide in soil water after forest clear-cutting. *Science* 233:867–869.

Bowman, M. J. 1977. Nutrient distribution and transport in Long Island Sound. *Estaurine Coastal Shelf Science* 5:531–548.

Bronk, D. A. 2002. Dynamics of organic nitrogen, pp. 153–249. In, Hansell, D. and C. A. Carlson (eds.), Biogeochemistry of Marine Dissolved Organic Matter. Academic Press, San Diego, CA.

Bronk, D. A. and P. M. Glibert. 1993. Application of a 15N tracer method to the study of dissolved organic nitrogen uptake during spring and summer in Chesapeake Bay. *Marine Biology* 115:501–508.

Bugna, G. C., J. P. Chanton, J. E. Cable, W. C. Burnett, and P. H. Cable. 1996. The importance of groundwater discharge to the methane budgets of nearshore and continental shelf waters in the northeastern Gulf of Mexico. *Geochimica et Cosmochimia Acta* 60:4735–4746.

Burt, R. A. 1993. Ground-water chemical evolution and diagenetic processes in the upper Floridan aquifer, southern South Carolina and northeast Georgia. U.S. Geological Survey Water-Supply Paper 2392, 76 p.

Bush, P. W. and R. H. Johnston. 1988. Ground-water hydraulics, regional flow, and ground-water development of the Floridan aquifer system in Florida and in parts of Georgia, South Carolina and Alabama. U.S. Geological Survey Professional Paper 1403-C, 88 p.

Butler, E. I., S. Knox, and M. I. Liddicoat. 1979. The relationship between inorganic and organic nutrients in sea water. *Journal of the Marine Biological Association* 59:239–250.

Cable, J. E., W. C. Burnett, and J. P. Chanton. 1997. Magnitudes and variations of groundwater seepage into shallow waters of the Gulf of Mexico. *Biogeochemistry* 38:189–205.

Cable, J. E., W. C. Burnett, J. P. Chanton, G. L. Weatherly. 1996. Estimating groundwater discharge into the northeastern Gulf of Mexico using radon-222. *Earth and Planetary Science Letters* 144:591–604.

Cai, W.-J., Y. Wang., J. Krest, and W. S. Moore. 2003. The geochemistry of dissolved inorganic carbon in a surficial groundwater aquifer in North Inlet, South Carolina, and the carbon fluxes to the coastal ocean. *Geochima et Cosmochima Acta* 67:631–639.

Cai, W.-J., W. J. Wiebe , Y. Wang, and J. E. Sheldon. 2000. The biogeochemistry of dissolved inorganic carbon and nitrogen in an intertidal marsh-estuarine complex in Southeastern U.S. *Limnology and Oceanography* 45:1743–1752.

Capone, D. G. and M. F. Bautista. 1985. A groundwater source of nitrate in nearshore marine sediments. *Nature* 313:214–216.

Capone, D. G. and J. M. Slater. 1990. Interannual patterns of water table height and groundwater derived nitrate in nearshore sediments. *Biogeochemistry* 10:277–288.

Cembella, A. D., N. J. Antia, and P. J. Harrison. 1984. The utilization of inorganic and organic phosphorous-compounds as nutrients by eukaryotic microalgae—a multidisciplinary perspective. *CRC Reviews in Microbiology* 10:317–391.

Chapelle, F. H. 1992. Ground-Water Microbiology and Geochemistry. Wiley, New York, NY. 468 p.

Chapelle, F. H. and P. B. McMahon. 1991. Geochemistry of dissolved inorganic carbon in a Coastal Plain aquifer. 1. Sulfate from confining beds as an oxidant in microbial CO_2 production. *Journal of Hydrology* 127:85–108.

Church, T. M. 1996. An underground route for the water cycle. *Nature* 380:579–580.

Corbett, D. R., J. P. Chanton, W. Burnett, K. Dillon, C. Rutkowski, and J. Fourqurean. 1999. Patterns of groundwater discharge into Florida Bay. *Limnology and Oceanography* 44:1045–1055.

Correll, D. L. 1981. Nutrient mass balance for the watershed, headwaters intertidal zone and basin of the Rhode River estuary. *Limnology and Oceanography* 26:1142–1149.

Correll, D. L., T. E. Jordan, and D. E. Weller. 1992. Nutrient flux in a landscape: Effects of coastal land use and terrestrial community mosaic on nutrient transport to coastal waters. *Estuaries* 15: 431–442.

Crandall, C. A. 1996. Shallow ground-water quality in agricultural areas of south-central Georgia, 1994. U.S. Geological Survey Water Resources-Investigations Report 96–4083. 23 p.

D'Elia, C. F., K. L. Webb, and J. W. Porter. 1981. Nitrate-rich groundwater inputs to Discovery Bay, Jamaica: A significant source of N to local reefs? *Bulletin of Marine Science* 3:903–910.

Dame, R. F., T. Chrzanowski, B. Kjerfve, H. McKellar, J. Spurrier, J. Vernberg, T. Williams, T. Wolaver, and R. Zingmark. 1991. Annual material processing by a salt marsh-estuarine basin in South Carolina. *Marine Ecology Progress Series* 72:153–166.

Dollar, S. J. and J. M. Atkinson. 1992. Effects of nutrient subsidies from groundwater to nearshore marine ecosystems off the island of Hawaii. *Estuarine and Coastal Shelf Science* 35:409–424.

Dosskey, M. G. and P. M. Bertsch. 1997. Transport of dissolved organic matter through a sandy forest soil. *Soil Society of America Journal* 61:920–927.

Dowdell, R. J., J. R. Burford, and R. Crees. 1979. Losses of nitrous oxide dissolved in drainage water from agricultural land. *Nature* 278:342–343.

Drever, J. I. 1982. The Geochemistry of Natural Waters: Surface and Groundwater Environments. Prentice-Hall, Englewood Cliffs, NJ. 436 p.

Earthinfo, Inc. 1997. Earthinfo CD^2. U.S. Geological Survey Water Quality. USGS Water Quality Bureau, Boulder, CO.

Fisher, T. R., J. Lawrence, W. Harding, D. W. Stanley, and L. G. Ward. 1988. Phytoplankton, nutrients, and turbidity in the Chesapeake, Delaware, and Hudson estuaries. *Estuarine and Coastal Shelf Science* 27:61–93.

Freeze, R. A. and J. A. Cherry. 1989. Groundwater. Prentice-Hall, Englewood Cliffs, NJ. 604 p.

Frimpter, M. H. and F. B. Gay. 1979. Chemical quality of groundwater of Cape Cod, Massachusetts. U.S. Geological Survey Water Resource Investigation. U.S. Department of the Interior, Boston, MA. 11 p.

Frink, C. R. 1991. Estimating nutrient exports to estuaries. *Journal of Environmental Quality* 20:717–724.

GESAMP. 1990. State of the marine environment. Rep. Stud. No. 39. Joint group of experts on the scientific aspects of marine pollution. United Nations Environment Program. 111 p.

Giblin, A. E. and A. G. Gaines. 1990. Nitrogen inputs to a marine embayment: the importance of groundwater. *Biogeochemistry* 100:109–128.

Gilliam, J. W., R. B. Daniels, and J. F. Lutz. 1974. Nitrogen content of shallow groundwater in the North Carolina coastal plain. *Journal of Environmental Quality* 3:147–151.

Girard, P. and C. Hillaire-Marcel. 1997. Determining the source of nitrate pollution in the Niger discontinuous aquifers using the natural 15N/14N ratio. *Journal of Hydrology* 199:239–251.

Grossman, E. L., B. K. Coffman, S. J. Fritz, and H. Wada. 1989. Bacterial production of methane and its influence on ground-water chemistry in east-central Texas aquifers. *Geology* 17:495–499.

Ham, L. K. 1997. Nutrients in surface water of the Georgia-Florida Coastal Plain, 1993–95. U.S. Geological Survey Water Resources-Investigations Report 96–4197. 8 p.

Harvey, J. W. and W. E. Odum. 1990. The influence of tidal marshes on upland groundwater discharge to estuaries. *Biogeochemistry* 10:217–236.

Hill, A. R. 1978. Factors affecting the export of nitrate-nitrogen from drainage basins in southern Ontario. *Water Research* 12:1045–1057.

Howarth, R. W. 1988. Nutrient limitation of net primary production in marine ecosystems. *Annual Review of Ecology and systhatics* 19:89–110.

Humborg, C., V. Ittekkot, A. Cociasu, and B. Vonbodungen. 1997. Effect of Danube River Dam on Black Sea biogeochemistry and ecosystem structure. *Nature* 386:385–388.

Hussain, N., T. M. Church, and G. Kim. 1999. Use of ^{222}Rn and ^{226}Ra to trace groundwater discharge into the Chesapeake Bay. *Marine Chemistry* 65:127–134.

Iqbal, M. Z., N. Krothe, and R. Spaulding. 1997. Nitrogen isotope indicators of seasonal source variability to groundwater. *Environmental Geology* 32:210–218.

Irvin, T. T., D. M. Bay, and L. E. Snipes. 1994. Georgia Agricultural Facts Book. Georgia Agricultural Statistics Service, Athens, GA.

Ittekkot, V. 1988. Global trends in the nature of organic matter in river suspensions. *Nature* 332:436–438.

Jackson, G. A. and P. M. Williams. 1985. Importance of dissolved organic nitrogen and phosphorous to biological nutrient cycling. *Deep-Sea Research* 32:223–235.

Johannes, Research E. 1980. The ecological significance of the submarine discharge of groundwater. *Marine Ecology Progress Series* 3: 365–373.

Jordan, T., D. Correll, and D. Weller. 1997a. Effects of agriculture on discharges of nutrients from coastal plain watersheds of Chesapeake Bay. *Journal of Environmental Quality* 26: 836–848.

Jordan, T., D. Correll, and D. Weller. 1997b. Nonpoint source discharges of nutrients from piedmont watersheds of Chesapeake Bay. *Journal of the American Water Resources Association* 33:631–645.

Jordan, T. and D. Weller. 1996. Human contributions to terrestrial nitrogen flux. *Bioscience* 46:655–664.

Karl, D. M. and G. Tien. 1997. Temporal variability in dissolved phosphorous concentrations in the subtropical North Pacific. *Marine Chemistry* 56:77–96.

Kay, E. A., L. S. Lau, E. D. Stroup, S. J. Dollar, D. P. Fellows, and R. H. F. Young. 1977. Hydrologic and ecological inventory of coastal waters of west Hawaii. Technical Report No. 105. Water Resources Center, University of Hawaii, Hondulu, HI. 93 p.

Kempe, S., M. Pettine, and G. Cauwet. 1991. Biogeochemistry of Europe rivers, pp. 169–211. In, Degens, E. T., S. Kempe, and J. E. Richey (eds.), Biogeochemistry of Major World Rivers, SCOPE 42. John Wiley and Sons, New York, NY.

Krest, J. M., W. S. Moore, and Rama. 1999. ^{226}Ra and ^{228}Ra in the mixing zones of the Mississippi and Atchafalaya Rivers: Indicators of groundwater input. *Marine Chemistry* 64: 129–152.

Krest, J. M., W. S. Moore, L. R. Gardner, and J. Morris. 2000. Marsh nutrient export supplied by groundwater discharge: Evidence from Ra measurements. *Global Biogeochemical Cycles* 14:167–176.

Kundell, J. E., J. Kealey, R. Klant, and L. Wilson. 1988. Management of Georgia's marshland under the Coastal Marshlands Protection Act of 1970. Carl Vinson Institute of Government, University of Georgia, Athens, GA. 34 p.

LaMontagne, M. G., R. Duran, and I. Valiela. 2003. Nitrous oxide sources and sinks in coastal aquifers and coupled estuarine receiving waters. *Science of the Total Environment* 309:139–149.

LaPointe, B. E. and J. O'Connell. 1989. Nutrient-enhanced growth of *Cladophora prolifera* in Harrington Sound, Bermuda: Eutrophication of a confined, phosphorus-limited marine ecosystem. *Estuarine and Coastal Shelf Science* 28:347–360.

LaPointe, B. E., J. D. O'Connell, and G. S. Garrett. 1990. Nutrient couplings between on-site sewage disposal systems, groundwater, and nearshore surface waters of the Florida Keys. *Biogeochemistry* 10:289–307.

LaRoche, J., R. Nuzzi, R. Waters, K. Wyman, P. G. Falkowski, and D. W. R. Wallace. 1997. Brown tide blooms in Long Island Sound's coastal waters linked to interannual variability in groundwater flow. *Global Change Biology* 3:397–410.

Lee, R. W. 1993. Geochemistry of ground water in the southeastern coastal plain aquifer system in Mississippi, Alabama, Georgia, and South Carolina. U.S. Geological Survey Professional Paper 1410-D. 72.p.

Leenheer, J. A., R. L. Malcolm, P. W. McKinley, and L. A. Eccles. 1974. Occurrence of dissolved organic carbon in selected ground-water samples in the United States. *Journal of Research of the U.S. Geological Survey* 2:361–369.

Legendre, L. O. and M. Gosselin. 1989. New production and export of organic matter to the deep ocean: Consequences of some recent discoveries. *Limnology and Oceanography* 34:1374–1380.

Li, L., D. A. Barry, F. Stagnitti, and J. Y. Parlange. 1999. Submarine groundwater discharge and associated chemical input to a coastal sea. *Water Resources Research* 35:3253–3259.

Lovley, D. R., F. H. Chapelle, and E. J. P. Phillips. 1990. Recovery of Fe(III)-reducing bacteria from deeply buried sediments of the Atlantic coastal plain. *Geology* 18:954–957.

Mallin, M. A., J. Burkholder, M. McIver, C. Shank, H. Glasgow, B. Touchette, and J. Springer. 1997. Comparative effects of poultry and swine waste lagoon spills on the quality of receiving streamwaters. *Journal of Environmental Quality* 26:1622–1631.

Malone, T. C., D. J. Conley, T. R. Fisher, P. M. Glibert, L. W. Harding, and K. G. Sellner. 1996. Scales of nutrient-limited phytoplankton productivity in Chesapeake Bay. *Estuaries* 19:371–385.

Marsh, J. A. 1977. Terrestrial inputs of nitrogen and phosphorous on fringing reefs of Guam. *Proceedings of the Second International Coral Reef Symposium* 2:332–336.

Matson, E. A. 1993. Nutrient flux through soils and aquifers to the coastal zone of Guam (Mariana Islands). *Limnology and Oceanography* 38:361–371.

Matson, P. M., W. Parton, A. Power, and M. Swift. 1997. Agricultural intensification and ecosystem properties. *Science* 277:504–509.

McClelland, J. W. and I. Valiela. 1998. Linking nitrogen in estuarine producers to land-derived sources. *Limnology and Oceanography* 43:577–585.

McClelland, J. W., I. Valiela, and R. H. Michener. 1997. Nitrogen-stable isotope signatures in estuarine food webs: A record of increasing urbanization in coastal watersheds. *Limnology and Oceanography* 42:930–937.

McCollister, T. 1993. Golf in Georgia. Longstreet Press, Atlanta, GA. 150 p.

Meade, R. H. and R. F. Vaccaro. 1971. Sewage disposal in Falmouth, MA. III. Predicted effects of inland disposal and sea outfall on groundwater. *Boston Society of Civil Engineering* 58:278–297.

Miller, J. A. 1990. Ground-water atlas of the United States: Alabama, Florida, Georgia, and South Carolina: U.S. Geological Survey Hydrologic Investigations Atlas 730-G. Atlanta, GA. 28 p.

Millham, N. P. and B. L. Howes. 1994. Freshwater flow into a coastal embayment: Groundwater and surface water inputs. *Limnology and Oceanography* 39:1928–1944.

Mook, W. G. and F. C. Tan. 1991. Stable carbon isotopes in rivers and estuaries, pp. 245–264. In, Degens, E. T., S. Kempe, and J. E. Richey (eds.), Biogeochemistry of Major World Rivers, SCOPE 42. John Wiley and Sons, New York, NY.

Moore, W. S. 1996. Large groundwater inputs to coastal waters revealed by [226]Ra enrichments. *Nature* 380:612–614.

Moore, W. S. 1999. The subterranean estuary: A reaction zone of groundwater and sea water. *Marine Chemistry* 65:111–126.

Moore, W. S., J. Krest, G. Taylor, E. Roggenstein, S. B. Joye, and R. Y. Lee, 2002. Thermal evidence of water exchange through a coastal aquifer: Implications for nutrient fluxes. *Geophysical Research Letters* doi-10: 1029/2002GL014923, 31 July 2002. p.

Mühlherr, I. H. and K. M. Hiscock. 1997. A preliminary assessment of nitrous oxide in Chalk groundwater in Cambridgeshire, U.K. *Applied Geochemistry* 12:797–802.

National Research Council. 2000. Clean Coastal Waters: Understanding and Reducing the Effects of Nutrient Pollution. National Academy Press, Washington, DC. 428 p.

Nearing, M. A., R. M. Risse, and L. F. Rogers. 1993. Estimating daily nutrient fluxes to a large Piedmont reservoir from limited tributary data. *Journal of Environmental Quality* 22:666–671.

Nixon, S. W. 1995. Coastal marine eutrophication: A definition, social causes, and future concerns. *Ophelia* 41:199–219.

Oberdorfer, J. A., M. A. Valentino, and S. V. Smith. 1990. Groundwater contribution to the nutrient budget of Tomales Bay, California. *Biogeochemistry* 10:199–216.

Orrett, K. and D. M. Karl. 1987. Dissolved organic phosphorous production in surface seawaters. *Limnology and Oceanography* 32:383–398.

Paerl, H. 1997. Coastal eutrophication and harmful algal blooms: Importance of atmospheric deposition and groundwater as "new" nitrogen and other nutrient sources. *Limnology and Oceanography* 42:1154–1165.

Paerl, H. W., J. L. Pinckney, J. M. Fear, and B. J. Peierls. 1998. Ecosystem responses to internal and watershed organic matter loading: consequences for hypoxia in the eutrophying Neuse River Estuary, North Carolina, USA. *Marine Ecology Progress Series* 166:17–25.

Pittman, J. P., H. H. Hatzell, and E. T. Oaksford. 1997. Spring contributions to water quantity and nitrate loads in the Suwannee River during base flow in July 1995. U.S. Geological Survey Water Resources Investigations Report 97–4152. 12 p.

Pomeroy, L. R., L. R. Shenton, R. D. H. Jones, and R. J. Reimold. 1972. Nutrient flux in estuaries. In, Nutrients and Eutrophication, *Limnology and Oceanography* Special Issue 1:274–291.

Portnoy, J. W., B. L. Nowicki, C. T. Roman, and D. W. Urish. 1998. The discharge of nitrate-contaminated groundwater from developed shoreline to marsh-fringed estuary. *Water Resources Research* 34:3095–3104.

Reay, W. G. and D. L. Gallagher. 1992. Groundwater discharge and its impact on surface water quality in a Chesapeake Bay inlet. *Water Resources Bulletin* 28: 1121–1134.

Rivkin, R. B. and E. Swift. 1985. Phosphorous metabolism of oceanic dinoflagellates: Phosphate uptake, chemical composition and growth of *Phyrocystis noctiluca*. *Marine Biology* 88:189–198.

Ronen, D., M. Magaritz, and E. Almon. 1988. Contaminated aquifers are a forgotten component of the global N_2O budget. *Nature* 335:57–59.

Rudy, M., K. McDonnell, I. Valiela, and K. Foreman. 1994. Dissolved organic nitrogen in groundwater bordering estuaries of Waquoit Bay, Massachusetts: Relations with watershed landscape mosaics. Biological Bulletin 187:278–279.

Ryther, J. H. and W. M. Dunston. 1971. Nitrogen, phosphorus and eutrophication in the coastal marine environment. *Science* 171:1008–1112.

Schultz, G. and C. Ruppel. 2002. Constraints on hydraulic parameters and implications for groundwater flux across the upland-estuary interface. *Journal of Hydrology* 260:255–269.

Seitzinger, S. and R. Sanders. 1997. Contribution of dissolved organic nitrogen from rivers to estuarine eutrophication. *Marine Ecology Progress Series* 159:1–12.

Sewell, P. L. 1982. Urban groundwater as a possible nutrient source for an estuarine benthic algal bloom. *Estuarine and Coastal Shelf Science* 15:569–576.

Shaw, T. J., W. S. Moore, J. Kloepfer, and M. A. Sochaski. 1998. The flux of Barium to the coastal waters of the southeastern United States: The importance of submarine groundwater discharge. *Geochimica et Cosmochimica Acta* 62:3047–3052.

Simmons, G. M. 1992. Importance of submarine groundwater discharge (SGWD) and seawater cycling to material flux across sediment/water interfaces in marine environments. *Marine Ecology Progress Series* 84:173–184.

Smith, J. O. 1988. Georgia and South Carolina stewardship of estuarine areas. Presented at Barrier Island/Salt Marsh Estuaries, Southeast Atlantic Coast: Issues, Resources, Status, and Management. Washington, DC, February 17, 1988.

Spalding, R. F. and J. D. Parrott. 1994. Shallow groundwater denitrification. *Science of the Total Environment* 141: 17–25.

Sprinkle, C. L. 1989. Geochemistry of the Floridan Aquifer System in Florida and in parts of Georgia, South Carolina and Alabama. U.S. Geological Survey Professional Paper 1403-I, 105 p.

Stanley, C. D., B. L. McNeal, P. R. Gilreath, J. F. Creighton, W. D. Graham, and G. Alverio. 1995. Nutrient loss trends for vegetable and citrus fields in west-central Florida. *Journal of Environmental Quality* 25:101–106.

Staver, K. W. and R. B. Brinsfield. 1996. Seepage of groundwater nitrate from a riparian agroecosystem into the Wye River Estuary. *Estuaries* 19:459–370.

Teal, J. M., A. E. Giblin, J. C. Goldman, D. G. Aubrey, and A. G. Gaines. 1982. The Coastal Impact of Groundwater Discharge: An Assessment of Anthropogenic Nitrogen Loading in Town Cove, Orleans, MA. Report Woods Hole Oceanographic Institution.

Ueda, S., N. Ogura, and E. Wada. 1991. Nitrogen stable isotope ratio of groundwater N_2O. *Geophysical Research Letters* 18: 1449–1452.

U.S. Census Bureau. 2000. Census 2000 redistricting data (P.L. 94–171).

Valiela, I. 1995. Marine Ecological Processes. 2nd ed. Springer-Verlag, New York, 686 p.

Valiela, I., J. L. Bowen, and K. D. Kroeger. 2002. Assessment of models for estimation of land-derived nitrogen loads to shallow estuaries. *Applied Geochemistry* 17:935–953.

Valiela, I., K. Foreman, M. LaMontagne, D. Hersh, J. Costa, P. Peckol, B. DeMeo-Anderson, C. D'Avanzo, M. Babione, C.-H. Sham, J. Brawley, and K. Lajtha. 1992. Coupling of watersheds and coastal waters: Sources and consequences of nutrient enrichment in Waquoit Bay, Massachusetts. *Estuaries* 15:443–457.

Valiela, I., P. Peckol, J. Costa, K. Foreman, J. M. Teal, B. Howes, and D. Aubrey. 1990. Transport of groundwater-borne nutrients from watersheds and their effects on coastal waters. *Biogeochemistry* 10:177–197.

Valiela, I., P. Peckol, C. D'Avanzo, K. Lajtha, W. Geyer, K. Foreman, D. Hersh, B. Seely, T. Isaji, and R. Crawford. 1997. Nitrogen loading from coastal watersheds to receiving estuaries: new method and application. *Ecological Applications* 7:358–380.

Valiela, I. and J. M. Teal. 1979. The nitrogen budget of a salt marsh ecosystem. *Nature* 280:652–656.

Valiela, I., J. M. Teal, S. Volkman, D. Shafer, and E. J. Carpenter. 1978. Nutrient and particulate fluxes in a salt marsh ecosystem: Tidal exchanges and inputs by precipitation and groundwater. *Limnology and Oceanography* 234:798–812.

Walling, D. E. and B. W. Webb. 1985. Estimating the discharge of contaminants to coastal waters by rivers: Some cautionary comments. *Marine Pollution Bulletin* 16:488–492.

Westerman, P. W., R. L. Huffman and J. S. Feng. 1995. Swine-lagoon seepage in sandy soil. *Transactions of the ASAE* 38(6):1749–1760.

Zekster, I. S. and H. A. Loaiciga. 1993. Groundwater fluxes in the global hydrologic cycle: Past, present and future. *Journal of Hydrology* 144: 405–427.

Zhang, C., E. L. Grossman, and J. W. Ammerman. 1997. Factors influencing methane distribution in Texas ground water. *Ground Water* 36:58–66.

8

Oxygen, Carbon Dioxide, and Estuarine Condition

Lawrence R. Pomeroy and Wei-Jun Cai

Summary by Lawrence R. Pomeroy and Daniel R. Hitchcock

Fish kills and obnoxious odors are symptoms of an estuary whose water is depleted of dissolved oxygen, a condition called hypoxia (or anoxia, when all of the oxygen is depleted). It is important that as communities develop in coastal watersheds, guidelines be developed for avoiding hypoxia and its consequences by detecting oxygen depletion and its causes before it has reached an extreme, and probably illegal, condition.

Although exceptions do occur, the water in healthy estuaries is usually nearly saturated with dissolved oxygen. Heterotrophs, organisms that do not photosynthesize, must extract dissolved oxygen from the water in order to respire and survive. All animals (including fishes) and many bacteria are heterotrophs. The process of respiration by estuarine organisms—taking in oxygen and giving off carbon dioxide—removes oxygen from the water. Most of the removal of oxygen from estuarine water is not by fishes but by naturally occurring and otherwise harmless microorganisms. They can grow on almost any source of organic matter that finds its way into the water. At the same time, oxygen is being added to the water by diffusion from the atmosphere across the surface of the estuary. Oxygen is also provided as a by-product of photosynthesis by microscopic algae (phytoplankton) in the water and by larger submerged algae and sea grasses. What is important is the maintenance of a balance between the inputs and the losses of oxygen. To encourage that balance, managers need to understand a few basic principles of estuarine structure and function, and the amount of dissolved oxygen or carbon dioxide in estuarine water must be measured.

In an estuary, some sources of organic matter that fuel respiration by microorganisms, depleting oxygen, are natural ones, such as those produced by the growth of plants in the water and in bordering wetlands. Organic matter from plants growing in the upland watershed of the estuary also enters the estuary with river water. While the algae and phytoplankton growing in the water are adding oxygen to the water as they grow, plants growing in uplands in the watershed and in bordering wetlands are not adding oxygen, since they exchange oxygen and carbon dioxide directly with the air. Those are natural sources of organic matter to estuaries that deplete the estuarine water of oxygen. Sometimes, even algae and phytoplankton can be a problem. If excessive fertilization of the estuary occurs as the result of inputs from sewage plants, farms, golf courses, or rain runoff from urban and suburban areas, phytoplankton growth in the estuary will increase. Much of that newly created excess organic material will then fall to the bottom of the estuary. The water near the bottom will lose all of its oxygen as microorganisms grow on the organic fallout, and that may lead to fish kills.

Since estuaries vary greatly in size, shape, depth, and the amount of fresh water passing through them, the processes affecting oxygen balance can also be quite different in their relative importance. In estuaries with substantial input of freshwater and a relatively deep central channel, the water is often layered (stratified), with saltier water near bottom and fresher water at the top. Even shallow estuaries may stratify thermally during summer, with a warm layer at the top and a cooler layer at the bottom. This layered structure of the water virtually seals off the bottom layer, making it more prone to oxygen depletion as organic matter falls out from the upper layer where the phytoplankton are growing. Estuaries as diverse as Chesapeake Bay and Long Island Sound may experience oxygen depletion near bottom because of the layered structure of the water. Any human-induced process that adds more organic matter to such estuaries increases the probability and extent of oxygen depletion near bottom. Each estuary must be understood in terms of its natural characteristics and the changes that it undergoes with the seasons. Potential human-introduced sources of organic matter or fertilizer must be identified and be related to the structure and function of that estuary in order to understand how it is likely to respond to those additions. Also, it is advisable to have in place at least a minimal program for routinely monitoring the distribution of dissolved oxygen at all water depths. Most estuaries today are probably receiving more of both organic matter and fertilizer than they did centuries ago. Environmental quality and resource managers should use all available clues to anticipate organic inputs that may overwhelm the estuary's ability to balance oxygen gains and losses. A new industry, housing development, or golf course in an estuary's watershed has a predictable outcome for water quality. Early signs of trouble should be documented with measurements of dissolved oxygen and action taken to prevent extreme loss of oxygen before it happens. Diffuse sources of organic matter and fertilizing nutrients, such as those from agriculture, septic tanks, suburban runoff, and golf courses, may be less obvious and more difficult to monitor than a sewage treatment plant. They are, however, frequent contributors to the organic matter and fertilizers that lead to oxygen depletion in the estuary.

So how does one measure the dissolved oxygen in an estuary? A variety of methods is now available, including simple dip-and-read devices that are easy to use. Their routine use can provide inexpensive monitoring of an estuary and can help to avoid expensive surprises. With such an approach, the most important considerations are when and where to take measurements. The basics include early-morning sampling and many measurements at all depths of water (technical details are provided in this chapter). If repeated surveys reveal changing oxygen concentrations or serious deficiencies of oxygen, or if potential legal considerations arise, dip-and-read monitoring may not be a sufficient approach. Precise methods for measuring dissolved oxygen require a considerable knowledge of chemistry, a substantial outlay of funds for equipment and supplies, and time-consuming work by technically competent personnel (detailed below). A prudent and proactive approach to the basic principle of achieving balance between the input and output of oxygen and carbon dioxide in an estuary provides the best insurance against undesirable outcomes, such as fish kills and noxious odors.

8.1 Introduction

Ecosystem respiratory processes and photosynthesis are key parameters of ecosystem condition that are characterized by their products: photosynthesis produces oxygen as a by-product, and aerobic respiration produces carbon dioxide. The importance of these parameters of the condition of aquatic ecosystems has been recognized for a century

(Delebecque 1898). The simple and robust Winkler (1888) method for measuring dissolved oxygen in water made it easy to detect waters of low oxygen content, which might be a threat to fishes and other aquatic life, and also to measure the potential biological oxygen demand of the organic matter in natural waters and sediments. These methods are frequently used in public health and sanitary engineering management. Carbon dioxide dissolved in water also can be measured, although this is somewhat more complex because of the reaction of carbon dioxide with water. Today, all of these measurements can be automated, simple dip-and-read instruments are available for measuring dissolved oxygen, and high-precision modifications that require careful calibration are also available. Moreover, we have improved our understanding of the location and magnitude of biological processes within estuaries, making predictions based on gas concentrations and gas exchanges more useful than ever. Understanding estuarine ecosystem structure and function, in addition to basic familiarity with methods, will help resource managers make good recommendations and decisions.

The concentration of dissolved oxygen and the partial pressure of carbon dioxide in estuarine water are valuable indicators of estuarine condition, because they are relatively simple to measure and we know what they mean. However, they must be used with some specific knowledge of the estuary, because some estuaries, or adjacent wetlands, may have naturally high respiratory rates (Cai and Wang 1998; Cai et al. 1999). This may also be true of certain freshwater streams and wetlands, some of which naturally fail to meet mandated dissolved oxygen minima (Hampton 1989).

In many states or countries, environmental protection agencies have adopted legally mandated criteria for the minimum permitted concentration of dissolved oxygen in natural waters. While this legalistic approach oversimplifies complex natural processes, sometimes leading to "illegal" oxygen concentrations owing to natural causes, it is a useful first line of defense against organic pollution and eutrophication. We will briefly discuss the processes that influence the concentrations of oxygen and carbon dioxide in estuarine waters and suggest why these easily monitored parameters are more useful than ever for protection of water quality and the safety of populations of estuarine organisms. Even in environments where we do not ordinarily expect to find depressed dissolved oxygen, the ease of measurement makes measuring it worthwhile insurance against unexpected impacts. In impacted environments, measuring dissolved oxygen is an inexpensive way to find trouble spots and to provide a focus that may help to minimize the need for more detailed and costly monitoring. We also review recent advances in methods of measuring dissolved oxygen and the carbon dioxide system in natural waters.

8.2 The Natural Metabolism of Estuaries

While the natural resource manager may not wish to investigate how natural aquatic food webs function, an awareness of what is currently known can be useful in understanding causes and solutions of management problems. For example, recent evidence suggests that most terrestrial organic matter of natural origins entering estuaries in river water either remains in estuarine sediments or is metabolized within the estuary, with essentially only the most refractory lignin phenols making their way into the sea (Benner and Opsahl 2001; Shi et al. 2001). This means that estuarine food webs are utilizing not only the products of local, estuarine photosynthesis but also organic matter from upstream, and, as a result, their net annual respiration often exceeds net annual photosynthesis (Goosen et al. 1997). If they are not re-aerated by wind and tidal mixing, such estuaries are vulnerable to anoxia owing to any additional anthropogenic loading of organic materials. Anthropogenic organic wastes impact numerous estuaries currently, and others have recovered following mitigation.

Excessive loading of nutrient elements, nitrogen, phosphorus, and silicon (i.e., eutrophication), stimulates high rates of photosynthesis in estuaries and, when the resulting organic matter decomposes, depletes the waters of oxygen and oversaturates them with carbon dioxide, initiating an increase in $p\mathrm{CO}_2$. The nutrients may originate from point sources, such as sewage-processing sites or animal feedlots, or they may originate from nonpoint sources, such as urban areas, golf courses, and the agricultural use of fertilizer. Nutrient loading sometimes results in summertime hypoxia in estuarine waters that is detrimental to fish and shellfish. So-called stratified estuaries, such as Chesapeake Bay, with a layer of low-salinity water capping a bottom layer of higher salinity, are especially vulnerable to hypoxia, because the bottom layer is not re-aerated continuously by winds and tides. Even well-mixed, shallow estuaries in the southeastern United States have experienced anoxia, but none in South Carolina or Georgia are currently considered to be eutrophic (Bricker et al. 1999).

8.2.1 Primary Production

The organic matter produced through photosynthesis, called primary production by ecologists, is the source of energy for nonphotosynthetic organisms in estuaries. This organic matter can originate from aquatic plants—phytoplankton, microalgae on the surfaces of rocks and sediments, and submerged macrophytes. It can also originate from the emergent vegetation in adjacent intertidal wetlands, and some of the organic matter produced by upland plants is transported into estuaries by rivers as particulate and dissolved organic matter, sometimes after residing in soils for years (Onstad et al. 2000). However, this latter "old" organic matter leached from soils is more resistant to further oxidation. From our perspective of the status of dissolved oxygen and carbon dioxide in estuaries, the origin of the organic matter is important. Submerged aquatic plants and phytoplankton are net producers of dissolved oxygen in the water, while emergent and upland vegetation is releasing oxygen to the atmosphere, not to the water. So, to the extent that emergent vegetation and terrestrial vegetation are the sources of estuarine organic matter, the eventual oxidation of that organic matter creates a net deficit in the balance of oxygen and carbon dioxide in the estuarine water that must be made up by gas exchanges across the surface of the water (Figure 8.1). Even in completely natural systems this does not always happen.

8.2.2 Estuarine Food Webs

To understand the budgets of oxygen and carbon dioxide in estuaries, we have to consider all living populations and all stocks of nonliving organic matter in inflowing river water, inflowing seawater, estuarine water, bottom sediments, adjacent tidal wetlands, and wetland sediments. In terms of oxygen demand and production of carbon dioxide, the smallest organisms are the most important. With each factor-of-ten decrease in body size, the metabolic rate per unit weight of organisms increases by a factor of approximately 1.75 (Peters 1983). This means that depending on nutrient availability and temperature (Pomeroy and Wiebe 2001), bacteria have from 10 to 1000 times the oxygen demand of an equal weight of fishes (Pomeroy 2001) (Figure 8.2). To find the oxygen demand in estuaries, we must measure the metabolic activity of microorganisms. Concentrations of bacteria tend to occur on organic detritus, and since organic detritus tends to sink, we can

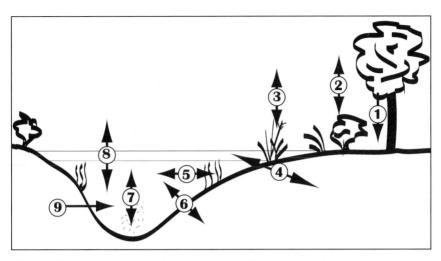

Fig. 8.1. Fluxes of oxygen and carbon dioxide that impact estuarine water.

(1) Respiratory degradation of terrestrial organic matter introduced with river water.
(2) Gas exchanges with the atmosphere by bordering terrestrial plants, with net production of organic matter that may be degraded in the estuary.
(3) Gas exchanges with the atmosphere by intertidal plants, with substantial fractions of organic production degraded in estuarine water or sediments.
(4) Complex aerobic and anaerobic degradation of organic matter in intertidal sediments with a net demand for oxygen and net production of carbon dioxide. Gas exchanges are with the atmosphere at all times by the intertidal macrophytes. Intertidal sediments exchange gases with the atmosphere during low tide and with the estuarine water during high tide.
(5) Gas exchanges of submerged macrophytes and benthic microalgae with estuarine water.
(6) Gas exchanges of bottom sediments and benthic macro- and microorganisms with estuarine water.
(7) Respiratory processes within the water column, primarily by aerobic microorganisms and largely associated with suspended or near-bottom organic particulate matter.
(8) Air-water gas exchanges influenced by wind waves and tidal currents.
(9) Inflow of ground water that may be enriched in carbon dioxide and depleted of oxygen.

expect a concentration of bacteria—and oxygen demand—in water along the bottom of the estuary, near the sediment surface, in near-surface sediments, and anywhere water currents concentrate sinking organic matter (see River Flow and Flushing Rate).

In the upper portions of some estuaries, the area of tidal wetlands exceeds the area of water at low tide, and these wetlands are usually populated with stands of highly productive emergent vegetation, such as *Spartina, Juncus, Phragmites,* or *Carex.* Since the vegetation is emergent, its photosynthesis is not adding oxygen to the water. Most

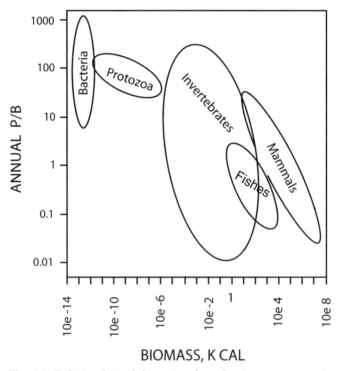

Fig. 8.2. Relationship of the ratio of production rate per unit biomass (P/B) to size of organisms demonstrates that even a small mass of bacteria and other microorganisms (Protozoa, Fungi) will outgrow and outrespire all other kinds of living organisms, making them the most influential in gas exchange in aquatic systems. Data from Banse and Mosher (1980); Ducklow (1992); Koslow (1996). Revised from Pomeroy (2001), with permission from the Institut de Scièncias del Mar.

wetland vegetation is not grazed by animals but instead dies in place, and a large fraction of the organic matter, especially that in roots and rhizomes, is incorporated into the intertidal sediments. This results in strong anaerobic bacterial respiration in the sediments, with a net release of carbon dioxide and demand for oxygen near the sediment surface. Water covering the wetlands at high tide thus tends to become depleted of oxygen and enriched in carbon dioxide, and when that water returns to the estuary at low tide and mixes into the estuary, the net oxygen concentration in the estuary is reduced and pCO_2 increased. Over a series of tidal cycles, the water in such estuarine zones can become greatly depleted in oxygen (Cai and Wang 1998; Cai et al. 1999). Because of temperature effects on bacterial metabolic rate, in temperate or subtropical estuaries this is primarily a summertime phenomenon (Shiah and Ducklow 1994a, b; Pomeroy et al. 2000). The net excess oxygen demand in wetlands is an entirely natural condition leading to naturally low oxygen concentrations in some tidal creeks and estuaries in summer (Ragotzkie 1959; Frankenberg 1975). This means that such estuaries will be highly sensitive to any anthropogenic processes that further reduce dissolved oxygen. It is important not only to monitor oxygen or carbon dioxide in such estuaries but also to predict changes in loading of organic matter from changing municipal or industrial sources that might quickly reduce oxygen concentrations to zero, with resultant damage to populations of fishes and invertebrates.

8.2.3 Outwelling

The discovery in the 1960s of the large amount of primary production of the emergent vegetation on intertidal wetlands led to the expectation that there must also be a significant export of organic matter from tidal wetlands to estuaries and through them to the sea, a process sometimes termed outwelling (Teal 1962; Odum 1968). Several investigators have failed to find observational evidence for a large flux of organic matter from tidal salt marshes (Haines 1977; Haines and Montague 1979; Nixon 1980; Chalmers et al. 1985), but the idea persists (Childers 1994; Childers et al. 2000), together with a recent assertion that regional differences can be seen, i.e., between marshes in western Europe and eastern North America (Heip et al. 1995; Dame and Allen 1996; Klap 1997). It is not necessary for us to recapitulate the debate here, but the reader should be aware that there is a debate; the outwelling concept remains alive but under continued scrutiny. The practical manager ought simply to go ahead and measure the local distributions of organic matter and oxygen and to act as the data indicate one should.

8.3 River Flow and Flushing Rate

Geomorphology and river flow rates interact to create a range of estuarine types having different dominant biological communities and different regimes of gas exchange. A number of classification schemes for estuaries have attempted to capture the significance of estuarine size, shape, depth, magnitude of river flow, and tidal effects (reviewed by Sheldon and Alber 2002). Deep or fjord-like estuaries and estuaries receiving major river flow usually have water that is stratified by salinity, with a brackish, river-dominated upper layer and a more saline bottom layer. Mixing between the layers at the halocline flushes salt back out to sea with the upper layer as it moves seaward, while the lower layer is resupplied from the sea, and there is a net movement of water in the lower layer from the sea toward the land. Because the lower layer is receiving organic fallout from the upper mixed layer, the lower layer often becomes partially or totally depleted of oxygen. Moreover, the landward movement of water in the lower layer transports particulate organic matter upstream, concentrating it at the head of the "salt wedge" where the last of the sea salt is mixed into the upper layer and transported back downstream. Such turbidity maxima are zones of high bacterial production and respiration (Baross et al. 1994). Oxygen is typically the most depleted in such systems in the upstream part of the lower water layer (Welsh et al. 1994). Although physical hydrographic structure sets the stage for the course of biological processes, it is the biology that is responsible for most carbon flux and transport (Kemp et al. 1997), but we must use knowledge of the physical and chemical structure of the estuary to lead us to potential trouble spots of hypoxia. Just as Jesse James said that he robbed banks "because that's where the money is," oxygen-depleting respiration, mostly by microorganisms, is going to occur where the labile organic matter is.

At the other morphological extreme from the fjord-estuaries, coastal lagoons with limited inputs of fresh water and very long flushing times may be thermally stratified in summer. In cases such as that of Long Island Sound, thermal stratification isolates a lower layer that becomes progressively depleted of oxygen during summer as a result of the oxygen demand by the benthos, in addition to that from sinking particulate organic matter produced by phytoplankton in the upper mixed layer (Welsh et al. 1994). Shallow coastal plain estuaries may be partially stratified by salinity or not at all stratified, depending on freshwater inflow, solar heating, tidal energy, and wind exposure. As noted above (Estuarine Food Webs), in spite of their lack of stratification, such estuaries may be depleted in oxygen during summer where some reaches are dominated by extensive intertidal marshes, and as a result those areas are sensitive to additional loading with organic carbon. This can be organic carbon introduced with sewage and other wastes, or it can result from additions of more nitrogen and phosphorus, which lead to more internal production of organic matter.

Wherever there is sea salt, there is movement of seawater as well as freshwater into the estuary. We think of the sea as being inherently salty but more dilute than estuaries with respect to its enrichment with nutrients (available forms of nitrogen and phosphorus) and labile organic matter. However, in locations where coastal upwellings impinge on the shore, the sea is the major, natural source of eutrophication. Examples of sea-enriched estuaries include Netarts Bay, Oregon, Walvis Bay, Namibia, and the *rías* of northwestern Spain. The only potential gas exchange problems in such systems of which we are aware are in Spanish *rías* where rafted mussel-culture leads to excessive deposition of organic matter (Blanton et al. 1987; Norueira et al. 1997).

8.4 Human Impacts on Estuarine Dissolved Gases

As we have outlined above, several types of estuaries are relatively sensitive to additional loading of organic matter because they are already partially depleted of dissolved oxygen, sometimes continuously and sometimes seasonally or over diel cycles. Good management of estuaries requires that we know the kind of estuary with which we are dealing and what one would expect its natural condition to be. In cases where anthropogenic impacts have occurred, understanding the estuarine types can help us to predict what we can expect once the impacts are mitigated. One cannot expect 100 percent saturation of oxygen everywhere in all circumstances. In some cases, natural oxygen concentrations may be near, or even below, legally mandated minima, and that is important to recognize (e.g., Hampton 1989). In such cases, of course, it is most important to minimize the human impacts.

8.4.1 Allochthonous Organic Matter

Many, if not most, of the world's estuaries are today receiving nutrients and organic matter originating from human activities that were not part of the historic river discharge, and we can never expect them to return to what were baseline conditions in the year 1500. The additional organic matter or nutrients may originate from industrial, municipal, agricultural, or forest-industry sources. The standing stocks of dissolved oxygen in estuaries reflect these changes in organic loading. Today, managers should be proactive by predicting expected gas concentrations in the water and initiating steps to make changes before large sections of an estuary become anoxic. Historically, many estuaries have become anoxic, especially during summer, when high temperature accelerates bacterial respiration of organic matter. Sometimes, the anoxia is not detected because it may occur only at night, when oxygen production by aquatic photosynthesis shuts down. Sampling very early in the morning, or installing automated, continuously recording devices, is advisable. Probably the most difficult problems for managers are nonpoint sources that may be difficult to identify and their effects difficult to quantify.

8.4.2 Eutrophication

More insidious than inputs of anthropogenic organic matter are inputs of forms of nitrogen and phosphorus that can be utilized by plants and which commonly result in blooms of phytoplankton. Phytoplankton blooms can lead to accumulation of organic matter in the lower layer of the water, in salt wedges, in deeper basins within an estuary, or in bottom sediments. The additional organic matter is metabolized and respired by microorganisms, especially in warm weather, producing an oxygen demand in addition to those already present. In estuaries that are naturally oxygen-deficient in summer, this can quickly lead to anoxia, with fish kills and other problems. Many inputs come from nonpoint sources, such as golf courses or agriculture, that can be difficult to identify and quantify, as well as difficult to regulate. Moreover, recovery may not have a linear relationship to mitigation but may occur only after some nutrient threshold has been passed (Smith et al. 1992). A classic example is Moriches Bay, Long Island, in which phytoplankton blooms developed during the 1950s that devastated its oyster industry, the source of the famous Blue Point oysters (Ryther 1954). The principal source of nutrients proved to be duck farms whose wastes were not controlled adequately. Thirty years later, with the duck farms long ago replaced by suburban housing developments, the blooms reappeared. This time, the principal sources of nutrients proved to be lawn fertilizers and dog feces. With urbanization, the eutrophication threshold was again surpassed and regulation had only become more difficult.

8.4.3 Impoundments

Impounding wetlands or uplands adjacent to estuaries can have two kinds of effects: (1) the effect of changing water circulation, estuarine water volume, flushing characteristics, and gas exchange of the impounded area, and (2) the effect of what is being done within the impoundment, which may at some point in time impact the estuary. Either of these may alter dissolved oxygen and carbon dioxide in estuarine waters. Since tidal wetlands can have a net oxygen demand, at least in summer (Cai et al. 1999; Pomeroy et al. 2000), impoundment might actually have a positive effect on dissolved oxygen and $p\text{CO}_2$ in estuarine water. However, this "benefit" could be offset by other important considerations, such as the loss of habitat for first-year shrimp, menhaden, and other harvestable species. The purpose of the impoundment and its consequences is the more serious issue. If the impoundment is intended for aquaculture, this will involve a concentration of the cultured organism and its food supplements that will be reflected in an accumulation of waste products and a resulting oxygen deficit. Some of either the organic wastes or the oxygen deficit is likely at some time to find its way into the estuary. With upland impoundments for the storage and oxidation of manure from animal feedlots, the problems are leakage and accidental rupture. Such an event on the Neuse River, North Carolina, undoubtedly affected dissolved oxygen, although that aspect of the spill was overshadowed by the induction of a dinoflagellate bloom (Burkholder et al. 1997).

8.5 Measurement Methods and Evaluation Strategies

8.5.1 The Winkler Method

Although most of the basic chemistry of the Winkler method has remained unchanged for a century, many improvements have been made in both the speed and the precision of the titration. Today, one ideally uses one of the several automatic titrators now commercially available (Oudot et al. 1988; Granéli and Granéli 1991). The automatic titrators not only speed the titration but also add precision. This can be further improved in a number of ways, either by titrating the entire contents of carefully calibrated bottles or by precision pipetting of the sample for titration (Pomeroy et al. 1994, Carnigan et al. 1998). Sensitivity as well as precision can be increased by statistical means through collection and titration of multiple water samples for each data point (Pomeroy et al. 1994). The speed of the automatic titration makes this feasible. A recently developed colorimetric method also achieves very high precision and avoids some of the manipulation problems of the precision titrations (Pai et al. 1993). Where less precision is needed or precision equipment is not available, Winkler samples can be read by a traditional burette technique (Carpenter 1965), or inexpensive dip-and-read oxygen electrodes can be taken into the field and oxygen concentrations read directly in the estuary. With any variation of the Winkler method, a number of substances that may be introduced into the water naturally or anthropogenically sometimes interfere and limit its accuracy. These are described by Clesceri et al. (1999).

8.5.2 Polarographic Oxygen Electrodes

In contrast to the Winkler method, which produces a single datum, polarographic oxygen electrodes (POEs) can produce either a single datum or a continuous record of change in pO_2, which might seem to make them ideal for measurements of respiratory rate. However, the potential advantage is partially offset by problems of drift and sensitivity to light or other interfering factors. They have nevertheless been widely used. A number of advanced models are now available, some of which are essentially free of drift (Langdon 1984; Griffith and Pomeroy 1995). POEs provide a clear choice for use in automated field data collection, and this has been done with considerable success (Kemp et al. 1994). They are also widely seen as the instrument of choice for rapid field surveys of dissolved oxygen. For this latter use, they have lower precision and sensitivity than Winkler samples but may be sufficiently precise for many purposes if well calibrated.

8.5.3 Electron Transport System

A method for estimating respiratory rates based on activity of the biochemical electron transport system (ETS) of all organisms, or all microorganisms, was developed by Packard (1971). Its use has been mostly in the ocean, where very low respiratory rates have made other methods difficult. It is currently quite popular in Europe. A continuing difficulty with the method has been calibration against some meaningful standard. Packard has recently made a new attempt to address this problem and has better defined the inherent problems in the method and potential ways to deal with them (Packard et al. 2004).

8.5.4 Measuring Dissolved Oxygen—General Considerations

The above methods do not measure exactly the same thing; the Winkler method measures total oxidizing equivalent, while POEs measure pO_2. As Pamatmat (1997) has argued, this difference can be a significant one if plankton or macrophytes are producing significant amounts of hydrogen peroxide as a metabolic by-product (which they do!), especially in respiratory rate measurements in oligotrophic waters in which enzymatic decay of H_2O_2 can produce what appears to be negative respiration, that is, oxygen production in the dark. The choice of method and the precision required should be considered during research planning in order to optimize what may be quick and simple measurements or may be a time-consuming and costly endeavor. Obviously, in systems in which large changes in dissolved oxygen occur, or when only large changes will be of concern to management, traditional chemical titrations (e.g., Carpenter 1965) will work, and simple, hand-held POE devices are rapid and may be sufficiently precise. Such approaches are not sufficient in oligotrophic systems or where small changes are of significance to management, or where peroxide production may be a significant interference. Nor are such devices usually sufficiently sensitive to measure respiratory rates in bottle experiments to determine respiratory rates or biological oxygen demand. The issues of precision and replicability can also, of course, become issues in legal proceedings.

It is important not to overlook questions of significance and replicability of dissolved oxygen measurements. Casual sampling is, unfortunately, just that. Many well-intended monitoring programs actually miss serious deficiencies in dissolved oxygen, either because the sampling times were limited to the middle of the day or because sampling simply was not sufficiently intense in a highly variable system (Taylor and Howes 1994). In addition to sampling intensity, Summers et al. (1997) also identify a significant sampling problem that they term "false positives," that is, more oxygen reported than really is present. This is obviously a question of faulty technique in either taking gas samples or reading them. Peroxide production might also be an issue here. It is especially tricky to handle samples that are very low in a dissolved atmospheric gas without introducing that gas to the samples from ambient air. Samples collected in bottles for later measurement of oxygen should be collected in a gas-tight sampler that is closed at depth. Oxygen samples should be taken from the sampler at once and drawn first, before any other samples (e.g., nutrients). The valve for withdrawing the samples should be equipped with a tube long enough to reach the bottom of the sample bottle, and the bottle should be overflowed from its bottom with a volume equal to three times the volume of the bottle, then sealed gas-tight, without bubbles. Reagents to chemically fix the dissolved oxygen should be added as soon as possible without adding air bubbles. If ordinary oxygen electrodes are used, they are inherently less precise than titration methods, and that makes frequent calibration of electrodes and their proper use an important consideration.

8.5.5 Measuring the Carbon Dioxide System

Four readily measurable parameters of the marine carbon dioxide system are pH, pCO_2, total dissolved inorganic carbon (DIC), and total alkalinity (TA). Carbon dioxide gas has a strong tendency to dissolve in water with a solubility of about 28 times that of oxygen (Weiss and Craig 1973; Weiss 1974; Cai and Wang 1998). Equilibrium between the gas phase and water is governed by a close approximation to Henry's Law, which states that for an ideally dilute solution, the gas vapor pressure of a volatile solute is proportional to its mole fraction in a solution (Cai and Wang 1998). Since CO_2 is the minor component of the marine inorganic carbon system (relative to bicarbonate), it is less sensitive to possible contamination by the atmosphere during sampling and analysis than is oxygen. But, like measurement of percent saturation of dissolved oxygen, pCO_2 is a useful indicator of the nature of biological processes in estuaries, and it can be measured directly or calculated from pH and DIC (Cai and Wang 1998). The pCO_2 of the estuarine water can then be compared with the atmospheric CO_2 concentration. The present average global atmospheric pCO_2 is approximately 370 microatmospheres. However, air in the coastal zone of developed countries is often higher owing to human influence (370–390 μatm, Cai unpublished data).

One method of measuring pCO_2 involves the combination of a gas-water equilibrator with an infrared CO_2 gas analyzer (Feely et al. 1998). Samples from the equilibrator can also be measured by gas chromatography. These systems tend to be large and complex and not suitable for unattended monitoring. Measurement of pCO_2 can also be performed spectrophotometrically using fiber optic sensors (DeGrandpre et al. 1993). This method can achieve a precision of 1 micro-atmosphere CO_2 (DeGrandpre et al. 1995) and a 99 percent response time of about 2 minutes (Wang et al. 2002, 2003). These sensors are mostly compact, have low power requirements, and can directly measure water-phase samples. They have been deployed successfully in mooring applications. Overall, however, handling pCO_2 is much more involved than that for oxygen sensors owing to the fact that pCO_2 sensors are not commercially available with full technical support.

In marine waters, DIC is most often measured by acidification of water samples and subsequent quantification of the extracted CO_2 gas by a coulometer or by an infrared CO_2 analyzer (Dickson and Goyet 1994; Cai and Wang 1998). The coulometric method is gradually being replaced by the infrared detector. The latter has the advantage of a small sample volume (0.1–1.0 ml), rapidity (<5 min/analysis), and high precision (standard deviation about 0.1 percent) (Cai and Wang 1998). The system is ideal for DIC analysis in sediment porewater, groundwater, culture incubation solutions, and coastal waters that require a wide range of concentrations and limited sample volume (Cai and Wang 1998; Cai et al. 1999; Cai and Reimers 2000; Cai et al. 2000). Salinity and H_2S do not interfere with the analysis.

8.6 Conclusion

Because of the basic importance of dissolved oxygen and carbon dioxide for aquatic life in estuaries, and the relative ease with which they can be measured in most situations, they should not be overlooked by those responsible for impact evaluations or estuarine monitoring. They are one of the most valuable of several first lines of defense against a range of pollution problems and one that usually can be carried out by personnel with limited training or experience, so long as a few simple rules that we have outlined are followed. Concentrations of these dissolved gases may be a guide to how and where to direct more intensive evaluations of other kinds. If done properly, these are measurements that can be counted on to "stand up in court" in situations where legal challenges must be met. We should remember and honor Winkler, who perceived the need and first devised a chemical method that permitted early understanding of the behavior of dissolved oxygen in natural waters and its potential significance.

Acknowledgment

Support for the research described here was provided in part by the NOAA Coastal Ocean Program under Grant NA960 PO113.

References

Banse, K. and S. Mosher. 1980. Adult body mass and annual production/ biomass relationships of field populations. *Ecological Monographs* 50:355–379.

Baross, J. A., B. Crump, and C. A. Simenstad (1994). Elevated "microbial loop" activities in the Columbia River estuarine turbidity maximum, pp 459–464. In, Dyer, K. R. and R. J. Orth (eds.), Changes in Fluxes in Estuaries: Implications from Science and Management. Olsen and Olsen, Fredensborg, Denmark.

Benner, R. and S. Opsahl. 2001. Molecular indicators of the sources and transformations of dissolved organic matter in the Mississippi River plume. *Organic Geochemistry* 32:597–611.

Blanton, J. O., K. R. Tenore, F. Castillejo, L. P. Atkinson, F. B. Schwing and A. Lavin. 1987. The relationship of upwelling to mussel production in the rias on the western coast of Spain. *Journal of Marine Research* 45:497–511.

Bricker, S. B., C. G. Clement, D. E. Pirhalla, S. P. Orlando and D. G. G. Farrow. 1999. National estuarine eutrophic assessment: Effects of nutrient enrichment in the nation's estuaries. National Ocean Service, National Oceanic and Atmospheric Administration, Silver Spring, MD. 83 p.

Burkholder, J. M., M. A. Mallin, H. B. J. Glasgow, L. M. Larsen, M. R. McIver, G. C. Shark, N. Deaner-Melia, D. S. Briley, J. Springer, B. W. Touchette, and E. K. Hannon. 1997. Impacts to a coastal river and estuary from rupture of a large swine waste holding lagoon. *Journal of Environmental Quality* 26:1451–1466.

Cai, W.-J. and Reimers, C. E. 2000. Sensors for in situ pH and pCO_2 measurements in seawater and at the sediment-water interface, pp. 75–119. In, Buffle, J. and G. Horvia (eds.), In-situ Monitoring of Aquatic Systems: Chemical Analysis and Speciation. Wiley, New York, NY.

Cai, W.-J., L. R. Pomeroy, M. A. Moran, and Y. Wang. 1999. Oxygen and carbon dioxide mass balance for the estuarine-intertidal marsh complex of five rivers in the southeastern U.S. *Limnology and Oceanography* 44:639–649.

Cai, W.-J. and Y. Wang. 1998. The chemistry, fluxes and sources of carbon dioxide in the estuarine waters of the Satilla and Altamaha rivers, Georgia. *Limnology and Oceanography* 43:657–668.

Cai, W.-J., W. J. Wiebe, Y. Wang, and J. E. Sheldon. 2000. Intertidal marsh as a source of dissolved inorganic carbon and a sink of nitrate in the Satilla river-estuarine complex in the southeastern U.S. *Limnology and Oceanography* 45:1743–1752.

Carnigan, R., A.-M. Blais, and C. Vis. 1998. Measurement of primary production and community respiration in oligotrophic lakes using the Winkler method. *Canadian Journal of Fisheries and Aquatic Science* 55:1078–1084.

Carpenter, J. H. 1965. The Chesapeake Bay Institute technique for the Winkler dissolved oxygen method. *Limnology and Oceanography* 10:141–143.

Chalmers, A. G., R. G. Wiegert, and P. L. Wolf. 1985. Carbon balance in a salt marsh: Interactions of diffusive export, tidal deposition and rainfall-caused erosion. *Estuarine, Coastal and Shelf Science* 21:757–771.

Childers, D. L. 1994. Fifteen years of marsh flumes: a review of marsh-water column interactions in southeastern USA estuaries, pp. 277–293. In, Mitsch, W. J. (ed.), Global Wetlands: Old World and New. Elsevier, Amsterdam, the Netherlands.

Childers, D. L., J. W. Day Jr., and H. N. McKellar Jr. 2000. Twenty more years of marsh and estuarine flux studies: revisiting Nixon (1980), pp. 391–423. In, Weinstein, M. P. and D. A. Kreeger (eds.), Concepts and Controversies in Tidal Marsh Ecology. Klewer, Dordrecht, the Netherlands.

Clesceri, L. S., A. E. Greenberg, and A.D. Eaton (eds.). 1999. Standard Methods for the Examination of Water and Wastewater, 20th ed. American Public Health Association, Washington, DC.

Dame, R. F. and D. M. Allen. 1996. Between estuaries and the sea. *Journal of Experimental Marine Biology and Ecology* 200:169–185.

DeGrandpre M. D. 1993. Measurement of seawater pCO_2 using a renewable-reagent fiber optic sensor with colorimetric detection. *Analytical Chemistry* 65:331–337.

DeGrandpre M. D., T. R. Hammar, S. P. Smith, and F. L. Sayles. 1995. In situ measurements of seawater pCO_2. *Limnology and Oceanography* 440:969–975.

Delebecque, A. 1898. Les Lacs Français. Chamerot et Renourard, Paris. 436 p.

Dickson, A. G. and C. Goyet 1994. DOE Handbook of Methods for the Analysis of the Various Parameters of the Carbon Dioxide System in Sea Water, Version 2. Carbon Dioxide Information Center, Oak Ridge, TN.

Ducklow, H. W. 1992. Factors regulating bottom-up control of bacterial biomass in open ocean plankton communities. *Archiv für Hydribiologie, Beiheft Ergebnisse Limnologie* 37:207–217.

Feely, R. A., R. Wanninkhof, H. B. Milburn, C. E. Cosca, M. Stapp, and P. P. Murphy. 1998. A new automated underway system for making high precision pCO_2 measurements on board research ships. *Analytica Chimica Acta* 377:185–191.

Frankenberg, D. 1975. Oxygen in a tidal river: Low tide concentration correlates linearly with location. *Estuarine and Coastal Marine Science* 4: 455–460.

Goosen, N. K., P. van Rijswijk, J. Kromkamp, and J. Peene. 1997. Regulation of annual variation in heterotrophic bacterial production in the Schelde estuary (SW Netherlands). *Aquatic Microbial Ecology* 12:223–232.

Granéli, A. and E. Granéli. 1991. Automatic potentiometric determination of dissolved oxygen. *Marine Biology* 108:343–348.

Griffith, P. C. and L. R. Pomeroy. 1995. Seasonal and spatial variations in pelagic community respiration on the southeastern U.S. continental shelf. *Continental Shelf Research* 15:815–825.

Haines, E. B. 1977. The origins of detritus in Georgia salt marshes. *Oikos* 29:254–260.

Haines, E. B. and C. L. Montague. 1979. Food sources of estuarine invertebrates analysed using $^{13}C/^{12}C$ ratios. *Ecology* 60:48–56.

Hampton, P. S. 1989. Dissolved-oxygen concentrations in a Florida wetlands stream, pp. 149–159. In, Fisk, D. W. (ed.), Wetlands: Concerns and Successes. American Water Resources Association, Bethesda MD.

Heip, C.H.R., N.K. Goosen, P.M.J. Herman, J. Kromkamp, J.J. Middelburg, and K. Soetaert. 1995. Production and consumption of biological particles in temperate tidal estuaries. *Oceanography and Marine Biology* 33:1–149.

Kemp, P. F., P. G. Falkowski, C. N. Flagg, W. C. Phoel, S. L. Smith, D. W. R. Wallace, and C. D. Wirick. 1994. Modeling vertical oxygen and carbon flux during stratified spring and summer conditions on the continental shelf, Middle Atlantic Bight, eastern U.S.A. *Deep-Sea Research II* 41:629–655.

Kemp, W. M., E. M. Smith, M. Marvin-DePasquale, and W. R. Boynton. 1997. Organic carbon balance and net ecosystem metabolism in Chesapeake Bay. *Marine Ecology Progress Series* 150:229–249.

Klap, V. A. 1997. Biogeochemical aspects of salt marsh exchange processes in the SW Netherlands. Ph.D. dissertation, Netherlands Institute of Ecology, University of Amsterdam, Amsterdam, Netherlands. 170 p.

Koslow, J. A. 1996. Energetic and life-history patterns of deep-sea benthic, benthopelagic and seamount-associated fish. *Journal of Fish Biology* 49:54–74.

Langdon, C. 1984. Dissolved oxygen monitoring system using a pulsed electrode: Design, performance and evaluation. *Deep-Sea Research* 31:1357–1367.

Nixon, S. W. 1980. Between coastal marshes and coastal waters — A review of twenty years of speculation and research on the role of salt marshes in estuarine productivity and water chemistry, pp. 437–525. In, Hamilton, P. and K. B. Macdonald (eds.), Estuarine and Wetland Processes. Plenum, New York, NY.

Norueira, E., F. F. Pérez, and A. F. Ríos. 1997. Seasonal patterns and long-term trends in an estuarine upwelling ecosystem (Ria de Vigo, N. W. Spain). *Estuarine, Coastal and Shelf Science* 44:285–300.

Odum, E. P. 1968. A research challenge: Evaluating the productivity of coastal and estuarine water. Proceedings of the Second Sea Grant Conference, University of Rhode Island, Newport, RI, pp. 63–64.

Onstad, G. D., D. E. Canfield, P. D. Quay and J. I. Hedges. 2000. Sources of particulate organic matter in rivers from the continental USA: Lignin phenol and stable carbon isotope compositions. *Geochimica et Cosmochimica Acta* 64:3529–3546.

Oudot, C., R. Gerard, P. Morin, I. Gningue. 1988. Precise shipboard determination of dissolved oxygen (Winkler procedure) for productivity studies with a commercial system. *Limnology and Oceanography* 33:146–150.

Packard, T. T. 1971. The measurement of respiratory electron-transport activity in marine phytoplankton. *Journal of Marine Research* 29:235–244.

Packard, T. T., D. Blasco, and M. Estrada. 2004. Modeling physiological processes in plankton on enzyme kinetic principles. *Scientia Marina* 68 (Suppl. 1):49–56.

Pai, S.-C., G.-C. Gong and K.-K. Liu 1993. Determination of dissolved oxygen in seawater by direct spectrophotometry of total iodine. *Marine Chemistry* 41:343–351.

Pamatmat, M. M. 1997. Non-photosynthetic oxygen production and non-respiratory oxygen uptake in the dark: a theory of oxygen dynamics in plankton communities. *Marine Biology* 129:735–746.

Peters, R. H. 1983. The Ecological Implications of Body Size. Cambridge University Press, Cambridge, UK. 329 p.

Pomeroy, L. R. 2001. Caught in the food web: Complexity made simple? *Scientia Marina* 65:31–40.

Pomeroy, L. R., J. E. Sheldon, and W. M. Sheldon Jr. 1994. Changes in bacterial numbers and leucine assimilation during estimations of microbial respiratory rates in seawater by the precision Winkler method. *Applied and Environmental Microbiology* 60:328–332.

Pomeroy, L. R., J. E. Sheldon, W. M. Sheldon Jr., J. O. Blanton, J. Amft, and F. Peters. 2000. Seasonal changes in microbial processes in estuarine and continental shelf waters of the south-eastern U.S.A. *Estuarine, Coastal and Shelf Science* 51:415–428.

Pomeroy, L. R. and W. J. Wiebe. 2001. Temperature and substrates as interactive limiting factors for marine heterotrophic bacteria. *Aquatic Microbial Ecology* 23:187–204.

Ragotzkie, R. A. 1959. Plankton productivity in estuarine waters of Georgia. *Publications in Marine Science, University of Texas* 6:147–158.

Ryther, J. H. 1954. The ecology of phytoplankton blooms in Moriches Bay and Great South Bay, Long Island, New York. *Biological Bulletin* 106:198–209.

Sheldon, J. E. and M. Alber. 2002. A comparison of residence time calculations using simple compartment models of the Altamaha river estuary, Georgia. *Estuaries* 25:1304–1317.

Shi, W., M.-Y. Sun, M. Molina, and R. E. Hodson. 2001. Variability in the distribution of lipid biomarkers and their molecular isotopic compositions in Altamaha estuarine sediments: implications for the relative contribution of organic matter from various sources. *Organic Geochemistry* 32:453–467.

Shiah, F.-K. and H. W. Ducklow. 1994a. Temperature and substrate regulation of bacterial abundance, production, and specific growth-rate in Chesapeake Bay, U.S.A. *Marine Ecology Progress Series* 103:297–308.

Shiah, F.-K. and H. W. Ducklow. 1994b. Temperature regulation of bacterioplankton abundance, production, and specific growth-rate in Chesapeake Bay. *Limnology and Oceanography* 39:1243–1258.

Smith, D. E., M. Leffler, and G. Mackiernan (eds.). 1992. Oxygen Dynamics in the Chesapeake Bay. Maryland Sea Grant College, College Park. 234 p.

Summers, J. K., S. B. Weisberg, A. F. Holland, J. Kou, V. D. Engle, and D. L. Breitberg. 1997. Characterizing dissolved oxygen conditions in estuarine environments. *Environmental Monitoring and Assessment* 45:319–328.

Taylor, C. D. and B. L. Howes. 1994. Effect of sampling frequency on measurement of seasonal primary production and oxygen status in near-shore coastal ecosystems. *Marine Ecology Progress Series* 108:193–203.

Teal, J. M. 1962. Energy flow in the salt marsh ecosystem of Georgia. *Ecology* 43:614–624.

Wang, Z., Y. Wang, and W.-J. Cai. 2003. A long pathlength liquid-core waveguide sensor for real-time pCO_2 measurements at sea. *Marine Chemistry* 84:73–84.

Wang, Z., Y. Wang, W.-J. Cai, and S. Y. Kiu. 2002. A long pathlength spectrophotometric pCO_2 sensor using a gas-permeable liquid-core waveguide. *Talanta* 57: 69–80.

Weiss, R. F. 1974. Carbon dioxide in water and seawater: The solution of a non-ideal gas. *Marine Chemistry* 2:203–215.

Weiss, R. F. and H. Craig. 1973. Precise shipboard determination of dissolved nitrogen, oxygen, argon, and total inorganic carbon by gas chromatography. Deep-Sea Research 20: 291–303.

Welsh, B. L., R. I. Welsh, and M. DiGiacomo-Cohen. 1994. Quantifying hypoxia and anoxia in Long Island Sound, pp. 131–137. In, Dyer, K. R. and R. J. Orth (eds.), Changes in Fluxes in Estuaries: Implications from Science and Management. Olsen and Olsen, Fredensborg, Denmark.

Winkler, L. W. 1888. Die Bestimmung des im Wasser gelösten Sauerstoffes. *Berichten Deutsche Chemische Gesellschaft* 21:2843–2855.

Part III

Contaminants and Their Effects

9

Chemical Contaminants Entering Estuaries in the South Atlantic Bight as a Result of Current and Past Land Use

Richard F. Lee and Keith A. Maruya

Summary by David A. Bryant

A survey of research related to toxic contamination in the South Atlantic Bight, a large, open embayment stretching roughly from Wilmington, North Carolina, to Jacksonville, Florida, reveals five major sources of contaminants that have the potential to harm coastal estuaries and the near-shore environment. All of these sources are associated with human activities within the watershed, some with historical activity, and some with relatively recent land use trends. The contaminants vary in the degree and manner in which they harm the environment and living (including human) populations. Their adverse effects depend on many factors, including how readily the compounds are broken down by light, water, or other processes, and how readily they are either accumulated or metabolized by organisms in the estuary.

Industry, largely in the form of pulp and paper plants, has discharged heavy metals and toxic organic contaminants into the region's coastal rivers for decades. These facilities are often located near urban areas. One survey found high concentrations of contaminants that persist in the environment (i.e., they degrade slowly), that accumulate in animals as they are transferred up the food chain, and that are toxic to the organisms that come in contact with them at locations in Brunswick, and Savannah, Georgia, and Charleston, South Carolina. Some of these plants continue to operate, and those that have closed have left a legacy of persistent-bioaccumulative-toxic (PBT) contaminants that remain in local sediment and continue to accumulate in the food chain.

Agriculture—cultivation of soybeans, corn, cotton, wheat, and peanuts—a historically important industry to the South Atlantic Bight coastal plain—employs a spectrum of insecticides, herbicides, and fungicides. The total use for 35 inventoried pesticides in the region ranges from 1,000 to 6,000 pounds per acre per year (1,100 to 6,700 kilograms per hectare per year). These pesticides can easily be washed into coastal waterways by rainfall and can contaminate the groundwater that seeps into rivers and streams. There are documented cases in the region of fish kills caused by pesticide runoff. However, a complex web of factors makes it difficult to predict the effects of pesticides in the environment. The chemical make-up of a pesticide; the properties of the soil on which it is applied; the method, amount, and timing of its application; atmospheric temperature; and the amount and type of rainfall can all determine whether it finds its way into local waterways.

Another important industry to the region is the planting and harvesting of slash and loblolly pine. This industry uses herbicides and insecticides to manage its crop throughout its 25- to 30-year rotation. The use of these chemicals is typically orders of magnitude less than in agriculture, and public pressure, economics, and changing techniques continue to reduce their use. However, that chemicals such as Permithrin, used to control the Nantucket tip moth, and the industry's most widely used herbicide, Triclopyr, continue to be found in sensitive waterways is cause for concern.

As the demand for retirement and second homes increases in the coastal Southeast, so does the number of golf courses. In Georgia, South Carolina, and North Carolina there are more than 150,000 acres of golf courses, with a disproportionate share on the coastal plain. A variety of pesticides are used to maintain golf courses, and repeated, heavy doses of these pesticides are often required to maintain greens and fairways. These pesticides can contaminate adjacent waterways or enter the system through groundwater. Many golf courses are located near sensitive estuarine environments.

Development in coastal areas can increase the toxic load significantly in estuaries. Pesticides may be overused by untrained consumers in suburban and urban settings, and rainfall can carry the excess application into nearby waterways. The pesticide Fipronil, used to control fire ants, is an example: it is often detected in suburban creeks and has been found to be extremely toxic to estuarine crustaceans, such as shrimp, at very low concentrations. No less troublesome is runoff from the roads and parking lots that proliferate as coastal communities grow. Rain washes heavy metals and hydrocarbons, such as motor oil, tire wear, and compounds in automobile exhaust, off these impervious surfaces and into local waterways, along with pesticides.

While industrial contaminants often enter the waterways directly, the fate of contaminants from other sources depends on a number of factors, such as local hydrology, the amount and frequency of precipitation, and the presence of buffers. Once contaminants have made their way to the estuary, another set of complex factors determines where they are distributed in the estuary and what organisms they effect. The toxicity of some contaminants can be reduced by water or light. However, many pesticides were created to resist such degradation, or their activities may actually be enhanced by water or sunlight. Some substances, such as mercury, accumulate readily in sediment and organisms, while others may be toxic to some organisms but harmless to others. For example, polycyclic aromatic hydrocarbons (PAHs), found in petroleum products, can be harmlessly metabolized by most fish but accumulate in the tissue of mollusks, such as oysters, clams, and mussels. While heavy doses of many contaminants can kill fish and other estuarine creatures, their nonlethal effect is often to inhibit growth and reproduction.

New guidelines for the selection, registration, and application of chemicals to the landscape are being developed to address pesticide problems in agriculture, silviculture, and golf course maintenance. Best management practices (BMPs) in these settings also suggest the use of vegetated buffers and retention ponds to prevent harmful chemicals from entering waterways. BMPs for coastal residential and commercial development attempt to constrain impervious surfaces and make use of retention ponds to capture contaminated runoff. Innovations in land use planning concentrate development in less-sensitive zones and mandate the use of buffers. As land use changes and more land is converted from agriculture and silviculture to residential and commercial uses, the challenge for the region will be to avoid the increased the load of toxins that typically come with roads, parking lots, and golf courses, and to prevent the critical changes to local hydrology caused by increased impervious surfaces.

9.1 Introduction

Wolfe (1992) suggested that the major inputs of contaminants into coastal waters are due to direct discharge from municipal and industrial outfalls, runoff from urban, agricultural, and silvicultural areas near the coast, and atmospheric fallout. Contaminants from these different inputs have been linked to effects on environmental quality, which can be assessed by various bioeffect indicators. The most useful bioeffect indicators are those linked to effects on the population, such as reproduction, growth, and survival. The focus of this paper is to describe the types, sources, fates, and potential for environmental impacts associated with chemical contaminants entering the South Atlantic Bight (SAB). Included are contaminants associated with current, past, and projected future land uses and responses of various bioeffect indicators to these contaminants.

Features of the coastal watersheds of the SAB include barrier islands, expansive marshes, dammed and undammed rivers with highly variable flows, and a broad continental shelf proximate to the Gulf Stream. Land along the coastal zones of the SAB has historically been used for agriculture and silviculture, with smaller pockets of urban, surburban, and resort development in such coastal cities as Wilmington, North Carolina, Charleston and Myrtle Beach, South Carolina, Savannah and Brunswick, Georgia, and Jacksonville, Florida. In a few cases, e.g., Charleston Harbor and Brunswick, there were large-scale point source discharges of potentially toxic heavy metals and organic contaminants. More recently, there has been extensive conversion of farm and silviculture lands to resort-type developments with low-density housing and golf courses. This review describes the nature, expected environmental behavior, and potential for toxic effects of chemical contaminants that enter SAB estuaries, including those resulting from direct discharge from municipal and industrial outfalls and runoff from urban, suburban, agricultural, and forested areas. Also noted are the sources and characteristics of several legacy contaminants at sites with severe contamination, some of which are or were designated as U.S. Environmental Protection Agency (U.S. EPA) Superfund sites.

9.2 Contaminant Sources

9.2.1 Industrial Contaminants

Timber, pulp and paper, and industrial chemical facilities producing chlorine and caustic compounds for pulp bleaching were early industries in the SAB coastal area (Sullivan 1997). Some of the contaminants left from these industries are classified as persistent, bioaccumulative, and toxic (PBT) (U.S. EPA 2003). Table 9.1 lists some sites in the SAB where high concentrations of PBT contaminants have been reported. Near Brunswick, Georgia, there was a chlor-alkali plant site with high sediment and biota concentrations of mercury and polychlorinated biphenyls. There is a second site with high sediment and biota concentrations of toxaphene, a chlorinated monoterpene formulation whose residues are a PBT contaminant of primary concern (U.S. EPA 2003).

9.2.2 Silviculture

Slash pine *(Pinus elliottii)* and loblolly pine *(Pinus taeda)* are planted and harvested by pulp and paper companies in many of the coastal areas of the SAB. Herbicides and insecticides are used to manage these pine forests. The sequence of pesticide application during a typical 25- to 30-year rotation is as follows: (1) postharvest site preparation; (2) pine release-midrotation; (3) herbaceous weed controls. Insecticides are applied as needed, usually during the first years of rotation when pine stands are being established (Fettig et al. 1998). The total amount of pesticides applied in silviculture on a total acreage basis is orders of magnitude less than in coastal agriculture (D. Moorhead, pers. comm.). A further reduction in pesticide application for forest management is continuing due to economics, changes in operations, and pressure from

Table 9.1. Industrialized Estuaries in the South Atlantic Bight with High Concentrations of PBT contaminants

Location	Contaminants	Reference
Charleston Harbor (SC)	Cr, Cu, Ni, Pb, Zn PAH	Sanger et al. (1999a,b) Hyland et al. (1996, 1998)
Purvis Creek (Brunswick, GA)	Hg, PCBs, PAHs	Kannan et al. (1997) Windom et al. (1976) Maruya and Lee (1998a,b)
Terry/Dupree Creek (Brunswick, GA)	Toxaphene	Maruya and Lee (1998a), Maruya et al. (2000)
Hutchinson Island (Savannah River, Savannah, GA)	Ag, Cd, Zn, Cr, Hg, Pb, PAHs, PCBs, DDT, dieldrin	Alexander et al. (1999)

the general public (McCormack 1994). One example of an operational alternative to pesticide application is the planting of optimum size pine seedlings, which results in greater survival and growth of slash pine relative to plots treated with the herbicide, imazapyr (South and Mitchell 1999).

Herbicides and insecticides commonly used in Georgia pine forests are listed in Table 9.2. One of the major insect pests of managed pine forest is the Nantucket tip moth (C. Fettig, pers. comm.). Permethrin and λ-cyalothrix applied 3–4 times annually at 0.21–0.65 g/hectare are generally used to control this insect (C. Fettig, pers. comm.). Imazapyr is used in combination with glyphosate, triclopyr, and picloram to control grasses, applied at 0.56–0.84 kg/ha (Harrington et al. 1998). Triclopyr is presently the most widely used herbicide in forest management (Larson et al. 1997).

Table 9.2. Pesticides Used in Silviculture in Coastal Georgia (C. Fettig, D. Moorhead, Personal Communications).

Chemical Name	Trade/Common name	Target
Insecticides		
acephate	Orthene	various insects
carbaryl	Sevin	various insects
λ-cyalothrix	Warrior T	tip moth
diflubenzuron	Dimilin	various insects
esfenvalerate	Asana	various insects
permethrin	Pounce	tip moth
spinosad		
Herbicides		
gyphosate	Accord	site preparation, weeds
hexazinone	Velpar	pine release, weeds
imazapyr	Arsenal	site preparation, pine release
picloram	Tordon	site preparation
sulfometuron methyl	Oust	site preparation, weeds
triclopyr	Garlon	site preparation, pine release

9.2.3 Agriculture

Agriculture, primarily soybeans, corn, wheat, and peanuts, is an important industry in SAB coastal areas. Total use for 35 inventoried pesticides ranges from 450 to 2,700 kilograms per year in the various estuarine drainage areas of this region (Pait et al. 1989, 1992; Table 9.3). A list of the herbicides, insecticides, and fungicides that are applied and their

Table 9.3. Properties of Pesticides Used in Coastal Areas of the South Atlantic Bight (from Pait et al. 1992).

Pesticide	Pesticide Class[a]	Water[a] Solubility (mg/L)	Soil Half-Life (days)	Toxicity to Fish LC50(mg/L)
Aciflurofen	H	VS	14	31
Alachor	H	242	30	4.3
Atrazine	H	33	130	2
Bensulide	H	25	140	0.32
Butylate	H	45	12	3
Carbaryl	I	30	8	1.6
Carbofuran	I	415	81	0.2
Chlorothalonil	F	0.6	45	0.032
Chlorpyrifos	I	0.4	30	0.14
Cyanazine	H	171	19	16
2,4-D	H	900	18	180
Diazinon	I	40	65	1.5
Disulfoton	I	25	9	0.52
Endosulfan	I	VIS	120	0.0015
Ethoprop	I	750	30	0.18
Fenvalerate	I	0.085	50	0.005
Fluometuron	H	90	30	43
Linuron	H	75	60	3
Malathion	I	145	11	0.32
Methamidophos	I	VS	3	34
Methyl Parathion	I	57	44	1.6
Metiram	F	VIS	60	32

(Continued)

Table 9.3. Properties of Pesticides Used in Coastal Areas of the South Atlantic Bight (from Pait et al. 1992). (*Continued*)

Pesticide	Pesticide Class[a]	Water[a] Solubility (mg/L)	Soil Half-Life (days)	Toxicity to Fish LC50(mg/L)
Metolachlor	H	530	40	8
Molinate	H	800	21	2
Parathion	I	24	6	0.036
PCNB	F	0.44	468	0.88
Permethrin	I	1	30	0.008
Phorate	I	50	25	0.0013
Profenofos	I	20	30	0.014
Propanil	H	225	15	18
Simazine	H	3.5	75	20
Terbufos	I	15	5	0.39
Thiobencarb	H	30	18	1.4
Trifluralin	H	1	81	930
Vermolate	H	90	11	2

[a] Abbreviations: H, herbicide; I, insecticide; F, fungicide; VS, very soluble; VIS, very insoluble; PCNB, pentachloronitrobenzene

application rates at 18 SAB coastal sites is given in Pait et al. (1989). For example, the combined Albermarle and Pamlico Sounds estuarine drainage area of coastal North Carolina and the Winyah Bay drainage area of coastal South Carolina had 1.8 million and 1.0 million kilograms of pesticide use per year, respectively. The major herbicides used were alachlor, atrazine, butylate, and dichlorophenoxyacetic acid (2,4-D); major insecticides were carbaryl, carbofuran, chlorpyrifos, and ethoprop; the major fungicide was chlorothalonil. In South Carolina, vegetable farms located adjacent to several estuaries receive applications of fenvalerate, azinphosmethyl, and endosulfan (Finley et al. 1999). Fish kills due to high concentrations of pesticides in runoff from agricultural fields have been documented in the SAB; the pesticides causing the fish kills included endosulfan, fenvalerate, and azinphosmethyl (Scott et al. 1987; Baughman et al. 1989; Trim and Marcus 1990; Scott et al., 1992; Ross et al. 1996). Analysis of pesticides in agricultural runoff associated with fish kills in Canada indicated that azinphos-methyl, chlorothalonil, and endosulfan exceeded fish 96 hour LC50 concentrations (Ernst et al. 2001). Some of the factors that affect pesticide runoff from agricultural fields include physicochemical properties of the pesticides, soil properties (grain size, organic content and quality, hydraulic conductivity), time interval between pesticide applications, ambient meterological conditions (temperature), and type and amount of rain (Willis and McDowell 1982).

9.2.4 Golf Courses

There are approximately 15,000 golf courses in the United States, with 300–400 new golf courses being developed each year (Ryals et al. 1998). Many of the new golf courses are located in SAB coastal areas. Using data from the National Golf Foundation, it has been calculated that there are 24,000 ha of golf course landscapes are in North Carolina alone, assuming that each 18-hole courses requires 53 ha, while there are approximately 34,000 ha of golf courses in Georgia and South Carolina (Balogh et al. 1992; Sadlon 1993). A wide variety of pesticides are used to maintain golf courses, including 2,4-D, chlorpyrifos, carbaryl, atrazine, chlorothalonil, metribuzinbentazone, simazine, and dimethyl tetra-chloroterephthalate (DCPA) (Balogh and Anderson 1992; Balogh et al. 1992; Miles et al. 1992; Ryals et al. 1998; Table 9.3). Golf course ponds in North Carolina were found to have high concentrations of chlorpyrifos (up to 60μg/L) after heavy summer rains (Ryals et al. 1998). Copper compounds are widely used to treat golf course lagoons to prevent algal blooms (Kamrin 1997). Different pesticides are used to maintain turf grass and putting greens (Balogh and Anderson 1992; Walker and Branham, 1992). Nationwide, Balogh and Anderson (1992) estimated that 5.5 million kilograms of pesticides were used in 1982 to maintain golf courses, with approximately 2 million kilograms used on golf courses in the southeastern region. Repeated, heavy doses of pesticides are sometimes needed to control certain pests that over time develop resistance to specific formulations (Balogh and Anderson 1992). For example, chlorothalonil is extensively and repeatedly used to control fungus on golf courses. Analyses of golf courses leachates in Georgia showed relatively high concentrations of the chlorothalonil degradation product, hydroxychlorothalonil, at concentrations ranging from 2.2 to 0.6 mg/L (Armbrust 2001).

9.2.5 Urban and Suburban Sources

Within urban and suburban environments there are daily uses of a variety of pesticides, many of which are also used in agriculture (Table 9.3). Examples of insecticides used in urban and suburban environments are chlorpyrifos, malathion, isazophos, isofenphos, pyrethroids (e.g., cypermethrin), and diazinon; herbicides used include diuron, 2,4-D, simazine, triclopyr, and oryzalin; fungicides used include chlorothalonil, fenarimol, benomyl, and captan (Racke 1993). A few pesticides have been developed specifically for urban and suburban use, such as isazophos, isofenphos, and oryzalin. Unfortunately, relatively few surveys of surface waters analyze for these compounds, which could allow a determination of the relative contribution of pesticides from urban, suburban, and agricultural areas to runoff. Fipronil, a phenylpyrazole insecticide, has been recently introduced to control fire ants, mole crickets, and termites in both golf courses and home gardens in the southeastern United States. Laboratory studies found that this compound was toxic at low concentrations (0.3 µg/L) to estuarine crustaceans (Fulton et al. 2002; Tingle et al. 2003; Wirth et al. 2004). Studies at a creek receiving runoff from a suburban development in South Carolina showed fipronil, chlorothalonil, and atrazine concentrations of 65, 13, and 500 ng/L (Lee and Fulton, unpublished data). All of these pesticides are widely used throughout SAB coastal areas (Larson et al. 1997).

Road runoff in suburban and urban environments contains many of the pesticides mentioned above as well as variety of aromatic compounds from motor oil, tire wear, and exhaust emissions. For example, where diuron was used to control weeds along a highway, the diuron concentration in runoff ranged from 0.1 to 10.8 µg/L (Huang et al. 2002). Herbicides frequently detected in urban and suburban creeks include prometon, simazine, atrazine, tebuthiuron, and metolachlor, while insecticides frequently detected in these creeks include diazinon, carbaryl, chlorpyrifos, and malathion (Bailey et al. 2000; Hoffman et al. 2000). High concentrations of polycyclic aromatic hydrocarbons (PAHs), benzthiazoles, and aromatic amines have been reported in sediments and waters receiving road runoff (Eganhouse et al. 1981; Bedding et al. 1982; Hoffman et al. 1984; Perry and McIntyre 1987; Spies et al. 1987; Hewitt and Rashed 1992; Stephensen et al. 2003). A study of highway runoff determined that the total loading rate of high molecular weight PAHs was 17 kg /km^2 of drainage area/y (Hoffman and

Quinn 1987). As an illustration of the influence of motor vehicle traffic and conversion of natural to impervious surface, Lee et al. (2004) found that PAH concentrations were extremely high (29 µg/g dry wt) in estuarine sediments nearest the roadway and adjacent parking lot in Hilton Head, South Carolina. Concentrations decreased rapidly with increasing distance from the roadway and parking lot (Table 9.4).

Table 9.4. Concentrations of PAHs (ng/g dry wt.) in Sediments from Stations at Different Distances from Highway Storm Drain in Hilton Head, South Carolina (from Lee et al. 2004).*

Compound	Station A	Station B	Station C
naphthalene	16	n.d	n.d.
2-methylnaphthalene	n.d	n.d	n.d.
1-methylnaphthalene	12	n.d	n.d.
biphenyl	n.d	n.d	n.d.
2,6-dimethylnaphthalene	n.d	n.d	n.d.
acenaphthylene	n.d.	n.d	n.d.
acenaphthene	151	n.d	n.d.
fluorene	220	n.d	n.d.
phenanthrene	3100	68	15
anthracene	530	15	n.d.
1-methylphenanthrene	155	16	n.d.
fluoranthene	4780	223	27
pyrene	3750	192	35
benz[a]anthracene	2350	92	14
chrysene	2250	110	14
benzo[b]fluoranthene	2620	169	16
benzo[k]fluoranthene	1290	83	n.d.
benzo[e]pyrene	1550	93	n.d.
benzo[a]pyrene	1930	93	n.d.
perylene	560	88	n.d.
indeno[1,2,3-cd]pyrene	1510	98	n.d.
dibenz[a,h]anthracene	550	39	n.d.
benzo[g,h,i]perylene	1500	102	n.d.
Total PAHs	28800	1480	120

*Tidal creek stations A, B, and C were located 1m, 100m, and 500m distances from a highway storm drain. N.d. = not detected (<10ng/g)

9.3 Fate of Contaminants in Estuaries

Fate is defined as the processes that transform, transport, modify, and/or sequester contaminants introduced into the environment, in this case coastal estuaries. Contaminants can be classified according to their environmental persistence, i.e., those with long half-lives (many months to years) are termed persistent or recalcitrant and those with short half-lives (days to a few months) are termed nonpersistent or labile. Such factors as aqueous solubility, vapor pressure, affinity for solid or biotic particles, and susceptibility to microbial degradation determine whether a contaminant will be persistent or labile (Table 9.5). Hydrophobic pesticides with large sediment-water distribution coefficients (K_d or K_{OC}) are attenuated via association with soil particles, while contaminants with high water solubility have higher

Table 9.5. Selected Organic Contaminants and Properties that Affect their Partitioning in Aquatic Ecosystems.*

Contaminant (class)+	C_{sol} (mg/L)	log K_{ow}	$t_{1/2}$ (days)	References
atrazine (H)	33	2.6	130	Pait et al. (1992) Gramatica & DiGuardo (2002)
alachlor (H)	242	3.5	30	Pait et al. (1992) Gramatica & DiGuardo (2002)
chlorothalonil (F)	0.8	2.9	0.2–8.8	Syngenta (2003) Gramatica & DiGuardo (2002)
permethrin (I)	0.04	3.5	90–120	Torstensson et al. (1999)
fenvalerate (I)	0.085	4.4	3	Southwick et al. (1995)
fluoranthene (PAH)	0.1	5.1	180–360	Verschueren (1983)
toxaphene (I)	0.4–3.3	4.8–6.4	1–3600	deGeus et al. (1999)
DDT (I)	0.003	6.2	730–5500	Verschueren (1983)

* C_{sol} = aqueous solubility; K_{ow} = octanol-water partition coefficient; $t_{1/2}$ = half-life
+ H = herbicide; I = insecticide; PAH = polycyclic aromatic hydrocarbon; F = fungicide

mobility. When contaminants enter upland and/or riparian areas, the fate of these contaminants is affected by local hydrology, precipitation, and evapotranspiration (Fares et al. 1996). In addition to physico-chemical properties, the mobility and transport of pesticides used in silviculture are controlled by hydrological processes such as surface runoff, infiltration, interflow, and leaching below the root zone, with highest concentrations found in waterways with little or no vegetative buffering (Neary et al. 1993). An important aspect of the fate of applied pesticides is the extent to which the pesticides enter surface waters through runoff and other effluents. When applied as a preemergent herbicide, 0.3 to 2 percent of the atrazine applied was lost to surface water (Larson et al. 1997). Two days after an application of fenvalerate to an agricultural field, the concentration of the compound in agricultural runoff was 1.9 µg/L (Southwick et al. 1995).

The biological fate of chemical contaminants is pertinent to the assessment of how coastal environmental quality changes with changing land use. Processes that influence the fate of contaminants once they enter the coastal ecosystem include uptake, elimination and excretion, biotransformation, and metabolism (Lee 2002; Mackay and Fraser 2000). Uptake may be passive (e.g., diffusion across membranes such as gills) or active (gut absorption of contaminants associated with food). Contaminants are lost from the biota by elimination and excretion pathways. Biotransformation is the modification of the compound by enzymatic processes, sometimes referred to as detoxification processes. Metabolism refers to breakdown of the chemical, with the products of this degradation used for cell energy needs. Contaminant bioavailability involves both biological and physiochemical processes that affect how contaminants are taken up into the biota. For persistent, hydrophobic contaminants, quantifying bioaccumulation and biomagnification processes is important to determining long-term biological impact. For labile, water soluble chemicals, bioaccumulation does not occur to a measurable extent, and thus such measures as lethal and sublethal effects indicate the bioavailability of the compounds.

9.3.1 Persistent Contaminants

Many synthetic organic and organometallic compounds such as PCBs, organochlorine pesticides (chlordane, DDTs, toxaphene), and organotins (tributyltin, triphenyltin) were synthesized for their stability and thus are not readily degraded. These compounds are hydrophobic, as indicated by their high log K_{ow}, and tend to partition into such lipid rich compartments as biota or sediments with high organic content (Table 9.5). Evidence of persistence was shown by the elevated concentrations of synthetic organochlorine compounds in sediment cores from the Savannah River estuary (Alexander et al. 1999). The high molecular weight PAHs, a common class of petroleum and pyrogenic contaminants in estuaries, are hydrophobic and persistent in anoxic sediments. However, they are readily metabolized by fish and many estuarine invertebrates, and thus most estuarine biota do not accumulate PAHs (Lee 2002). Exceptions are bivalve mollusks, e.g., mussels, clams, and oysters, which appear to lack an effective PAH metabolism system, and thus these animals accumulate both PAHs and other persistent contaminants. Quantitative structure activity relationships (QSARs) between physiochemical properties (e.g., K_{ow}), environmental persistence, bioaccumulation, and trophic transfer have been used to predict the generalized environmental behavior of various classes of hydrophobic contaminants in mostly freshwater and polar ecosystems (Kidd et al. 1995; Fisk et al. 1998). Very few studies have addressed the behavior and fate of persistent contaminants in complex coastal and estuarine ecosystems (Maruya and Lee 1998b).

9.3.2 Nonpersistent Contaminants

Because nonpersistent contaminants are generally soluble in water, they are susceptible to degradation by both abiotic and biotic processes. For example, pyrethroid insecticides, such as permethrin and fenvalerate, were found to be highly degradable in simulated aquatic ecosystems, with half-lives ranging from 0.5 to 31 days at 15–19°C (Lutnicka et al. 1999). For organochlorine compounds the rate of biotransformation depends on both the degree of chlorination as well as the stereochemistry, i.e., chlorine substitution patterns. Chlorinated biphenyls with 1 or 2 chlorines are rapidly degraded under aerobic conditions (Brown et al. 1987; Harkness et al. 1993) while PCB congeners with more than 6 chlorine atoms are only very slowly transformed (Palekar et al. 2003). Widely used organochlorine pesticides such as atrazine, fenvalerate, trichloropyr, chlorothalonil, and endosulfan have 1, 1, 3, 4, and 6 chlorines per molecule, respectively. Half-lives of these pesticides in soil range from 45 to 130 days (Table 9.3). Many of the organic contaminants of concern, e.g., PAHs and many pesticides, contain aromatic rings that are subject to microbial degradation in aerobic waters and sediments by metabolic pathways involving ring oxidation and ring cleavage (Crosby 1998). As a result of bioturbation, sediment-dwelling fauna, such as meiobenthos, can also biotransform sediment-associated organic contaminants (Forbes et al. 1996; Christensen et al. 2002; Selck et al. 2003). Gardner et al. (1979) found that without the presence of the polychaete, *Capitella capitata*, in an estuarine sediment, benzo(a)pyrene was not degraded during a 60-day study. In addition to biological degradation, abiotic mechanisms such as photolysis, oxidation, reduction, and hydrolysis may transform a variety of pesticides and PAHs (Zepp et al. 1985; Hwang et al. 1986; Crosby 1998). For example, organic rich clays can catalyze the hydrolysis of organophosphate pesticides (Soma and Soma 1989). The half-life of chlorothalonil in estuarine water was determined to be eight to nine days (Walker et al. 1988). This rapid degradation is primarily due to photolysis, which degrades chlorothalonil in natural waters to hydroxychlorothalonil (half-life = 57 days) (Armbrust 2001). Because of the possibility of many transformation processes, simple models that can adequately predict environmental fate for nonpersistent contaminants are rare.

9.4 Conclusions

A variety of chemical contaminants from a variety of sources can enter surface and groundwaters of coastal ecosystems of the SAB. Depending on their mobility and rates of degradation, these contaminants can then enter the estuaries that are such a common feature of this region. While fish kills can result when very high concentrations of pesticides enter the runoff from agricultural, urban, or suburban environments, the most likely effects of contaminants in SAB estuaries are sublethal, which affect such parameters as growth and reproduction of estuarine animals. Effects on growth and reproduction are linked to effects on the population dynamics of estuarine animals. The land use changes taking place in the SAB are leading to changes in both types and quantities of various contaminants. The rapid growth of the suburban and urban areas in this coastal environment will increase the percentage of impervious surface and later the hydrology of previously undisturbed landscapes, resulting in increased loading of contaminants associated with runoff and diverted water flows from golf courses, roads, and urban and suburban areas over the next several decades.

Several strategies to reduce and/or minimize the impact of contaminants released due to increasing development and land use change are available, including the design and implementation of Best Management Practices (BMPs). One such practice that deals with contaminants from nonpoint sources is the use of retention ponds located between the contaminant sources and the receiving estuary. Properly designed and operated retention ponds can reduce contaminant loading by allowing contaminant-bearing suspended sediments to settle and for the more labile contaminants to degrade to less toxic forms before entering the estuaries. In concert with the implementation of retention ponds are periodic maintenance and removal of contaminated pond sediments. Alternatively, the mobility and leaching potential of potentially toxic contaminants associated with bedded pond sediments could be reduced by in situ sequestration and/or vitrification technology. Land use planning strategies include minimizing impervious surface conversion, concentrating development in less sensitive area or in areas with integrated wastewater/runoff treatment facilities, and mandating effective minimum vegetated buffer zones. Such strategies would reduce contaminant loading by reducing runoff volume. Coincidentally, soil recharge and infiltration times would increase, and highly contaminated wastewater and runoff would be subject to engineered treatment processes. A final strategy would address the selection, registration, and application of the chemicals. For example, the use of less persistent and toxic chemicals will decrease the risk of in situ effects on estuarine biota. Reducing the total mass of applied chemicals, particularly pesticides used in silviculture, agriculture, and turf and garden management to a lower level will reduce the total inventory of contaminants in SAB coastal estuaries.

Acknowledgments

The support to prepare this review was provided by the NOAA Center for Coastal Ocean Research/Coastal Ocean Program through the South Carolina Sea Grant Consortium pursuant to National Oceanic and Atmospheric Administration Award number NA960PO113 is acknowledged.

References

Alexander, C., R. Smith, B. Loganathan, J. Ertel, H. L. Windom, and R. F. Lee. 1999. Pollution history of the Savannah River estuary and comparisons with Baltic Sea pollution history. *Limnologica* 29:267–273.

Armbrust, K. L. 2001. Photodegradation of hydroxychlorothalonil in aqueous solutions. *Environmental Toxicology and Chemistry* 20:2699–2703.

Baughman, D.S., D. W. Moore D, and G. L. Scott. 1989. A comparison and evaluation of field and laboratory toxicity tests with fenvalerate on an estuarine crustacean. *Environmental Toxicology and Chemistry* 8:417–429.

Bailey, H. C., L. Deanovic, E. Reyes, T. Kimball, K Larson, K. Cortright, V. Connor, and D. E. Hinton. 2000. Diazinon and chlorpyrifos in urban waterways in northern California. *Environmental Toxicology and Chemistry* 19: 82–87.

Balogh, J. C. and J. L. Anderson. 1992. Environmental impacts of turfgrass pesticides, pp. 221–353. In, Balogh J. C. and W. J. Walker (eds.) Golf Course Management and Construction—Environmental Issues. Lewis, Boca Raton, FL.

Balogh, J. C., V. A. Gibeault, W. J. Walker, M. P. Kenna, and J. T. Snow. 1992. Background and overview of environmental issues, pp. 1–37. In, Balogh, J. C. and W. J. Walker (eds.), Golf Course Management and Construction—Environmental Issues. Lewis, Boca Raton, FL.

Bedding, N. D., A. E. McIntyre, R. Perry, and J. N. Lester. 1982. Organic contaminants in the aquatic environment. I. Sources and occurrence. *Science and Total Environment* 25:143–167.

Brown, J. F., D. L. Bedard, M. J. Brennan, J. C. Carnahan, H. Feng, and R. E. Wagner. 1987. Polychlorinated biphenyl dechlorination aquatic sediments. *Science* 236:709–712.

Christensen, M., G. T. Banta, and O. Andersen. 2002. Effects of the polychaetes *Nereis diversicolor* and *Arenicola marina* on the fate of and distribution of pyrene in sediments. *Marine Ecology Progress Series* 237:159–172.

Crosby, D. G. 1998. Environmental Toxicology and Chemistry. Oxford University Press, New York, NY. 336 p.

de Geus, H. J., H. Besselink, A. Brouwer, J. Klungsøyr, B. McHugh, E. Nixon, G. G. Rimkus, P. G. Wester, and J. deBoer. 1999. Environmental occurrence analysis and toxicology of toxaphene compounds. *Environmental Health Perspectives* 107 (Suppl 1):115–144.

Eganhouse, R. P., B. R. T. Simoneit, and I. R. Kaplan, 1981. Extractable organic matter in urban stormwater runoff. 2. Molecular characterization. *Environmental Science and Technology* 15:315–326.

Ernst, B., K. Koe, P. Jackman, and J. Mutch. 2001. Diagnosing pesticide-induced fish kills in streams draining agricultural areas. *SETAC Globe* 2:20.

Fares, A., R. S. Mansell, and N. B. Comerford. 1996. Hydrological aspects of cypress wetlands in coastal region pine forests and impacts of management practices. *Proceedings of the Florida Soil Crop Science Society* 55:52–58.

Fettig, C. S., C. W. Berisford, and M. J. Dalusky. 1998. Revision of a timing model for chemical control of the Nantucket pine tip moth (Lepidoptera: Tortricidae) in the southeastern coastal plain. *Journal of Entomological Science* 33:336–342.

Finley, D. B., G. I. Scott, J. W. Daugomach, S. L. Layman, L. A. Reed, M. Sanders, S. K. Siversten, and E. D. Strozier. 1999. An ecotoxicological assessment of urban and agricultural nonpoint source runoff effects on the grass shrimp, *Palaemonetes pugio*, pp. 143–173. In, Lewis, M. A., R. L. Powell, M. K. Nelson, M. G. Henry, S. J. Klaine, G. W. Dickson, and F. L. Mayer (eds.), Ecological Risk Assessment for Wetlands. SETAC Press, Pensacola, FL.

Fisk, A. T., R. J. Norstrom, C. D. Cymbalisty, and D .C. G. Muir. 1998. Dietary accumulation and depuration of hydrophobic organochlorines: bioaccumulation parameters and their relationship with the octanol/water partition coefficient. *Environmental Toxicology and Chemistry* 17:951–961.

Forbes, V. E., T. L. Forbes, and M. Holmer. 1996. Inducible metabolism of fluoranthene by the opportunistic polychaete *Capitella* sp. I. *Marine Ecology Progress Series* 132:63–70.

Fulton, M. H., P. Pennington, M. DeLorenzo, E. Wirth, P. Key, D. Bearden, S. Siversten B. Shaddrix, S. Walse, and J. Ferry. 2002. Effects of fipronil in an estuarine mesocosm, p. 133. In, Abstracts of SETAC 23rd Annual Meeting in North America, November 16–20, 2002, Society of Environmental Toxicology and Chemistry, Pensacola, FL.

Gardner, W. S., R. F. Lee, K. R. Tenore, and L. W. Smith. 1979. Degradation of selected polycyclic aromatic hydrocarbons in coastal sediments: importance of microbes and polychaete worms. *Water, Air and Soil Pollution* 11:339–347.

Gramatica P. and A. DiGuardo. 2002. Screening of pesticides for environmental partitioning tendency. *Chemosphere* 47:947–956.

Harkness, M. R., J. B. McDermott, D. A. Abramowicz, J. J. Slavo, W. P. Flanagan, M. L. Stephens, F. J. Mondello, R. J. May, J. H. Lobos, K. M. Carroll, M. J. Brennan, A. A. Bracco, K. M. Fish, G. L. Warner, P. R. Wilson, D. K. Dietrich, D. T. Lin, C. B. Morgan, and W .L. Gately. 1993. In situ stimulation of aerobic PCB biodegradation in Hudson River sediments. *Science* 259:503–507.

Harrington, T. B., P. J. Minogue, D. K. Lauer, and A. Q. Ezell. 1998. Two year development of southern pine seedlings and associated vegetation following spray and burn site preparation with imazapyr alone or in mixture with other herbicides. *New Forests* 15:89–106.

Hewitt, C. N. and M. B. Rashed. 1992. Removal rates of selected pollutants in the runoff waters from a major rural highway. *Water Research* 26:311–319.

Hoffman, E. J. and J. G. Quinn. 1987. Chronic hydrocarbon discharges into aquatic environments: II-Urban runoff and combined sewer overflows, pp. 97–113. In, Vandermeulen, J. H. and S. E. Hrudey (eds.,) Oil in Freshwater: Chemistry, Biology, Countermeasure Technology. Pergamon, New York, NY.

Hoffman, E. J., G. L. Mills, J. S. Latimer, and J. G. Quinn. 1984. Urban runoff as a source of polycyclic aromatic hydrocarbons to coastal waters. *Environmental Science and Technology* 18:580–587.

Hoffman, R. S., P. D. Capel, and S. J. Larson. 2000. Comparison of pesticides in eight U.S. urban streams. *Environmental Toxicology and Chemistry* 19:2249–2258.

Huang, X., M. Fischer, R. White, T. Pedersen, S. Given, Y. Lu, and T. Young. 2002. Diuron runoff and treatment from highway roadside, p. 331. In, Abstracts of SETAC 23rd Annual Meeting in North America, November 16–20, 2002, Society of Environmental Toxicology and Chemistry, Pensacola, FL.

Hwang, H.-M., R. E. Hodson, and R. F. Lee. 1986. Degradation of phenol and chlorophenols by sunlight and microbes in estuarine water. *Environmental Science and Technology* 20:1002–1007.

Hyland, J. L., T. J. Herrlinger, T. R. Snoots, A. H. Ringwood, R. F. Van Dolah, C. T. Hackney, G. A. Nelson, J. S. Rosen, and S. A. Kokkinakis. 1996. Environmental quality of estuaries of the Carolinian Province: 1994. NOAA Technical Memorandum NOS ORCA 97, National Oceanic and Atmospheric Administration, Rockville, MD.

Hyland, J., T. Snoots, and L. Balthis. 1998. Sediment quality of estuaries in the southeastern U.S. *Environmental Monitoring and Assessment* 51:331–343.

Kamrin, M.A. 1997. Pesticide Profiles. Lewis, Boca Raton, FL.

Kannan, K., K. A. Maruya, and S. Tanabe. 1997. Distribution and characterization of polychlorinated biphenyl congeners in soils and sediments from a Superfund site contaminated with Aroclor 1268. *Environmental Science and Technology* 31:1483–1488.

Kidd, K. A., D. W. Schindler, R. H. Hesslein, and D. C. G. Muir. 1995. Correlation between stable nitrogen isotope ratios and concentrations or organochlorines in biota from a freshwater food web. *Science of the Total Environment* 101/161:381–390.

Larson, S. J., P. D. Capel, and M. S. Majeewski. 1997. Pesticides in Surface Waters: Distribution, Trends, and Governing Factors. Ann Arbor Press, Chelsea, MI. 373 p.

Lee, R. F. 2002. Bioavailability, biotransformation, and fate of organic contaminants in estuarine animals, pp. 97–126. In, Newman, M. C., M. R. Roberts, and R. C. Hale (eds.), Coastal and Estuarine Risk Assessment. Lewis, Boca Raton, FL.

Lee, R. F., K. A. Maruya, and K. Bulski. 2004. Exposure of grass shrimp to sediments receiving highway runoff: effects on reproduction and DNA. *Marine Environmental Research* 58:713–717.

Lutnicka, H., T. Bogacka, and L. Wolska. 1999. Degradation of pyrethroids in an aquatic ecosystem model. *Water Research* 33:3441–3446.

Mackay, D. and A. Fraser. 2000. Bioaccumulation of persistent organic chemicals: mechanisms and models. *Environmental Pollution* 110:375–391.

Maruya, K. A. and R. F. Lee. 1998a. Arochlor 1268 and toxaphene in fish from a southeastern estuary. *Environmental Science and Technology* 32:1069–1075.

Maruya, K. A. and R. F. Lee. 1998b. Biota-sediment accumulation and trophic transfer factors for extremely hydrophobic polychlorinated biphenyls. *Environmental Toxicology and Chemistry* 17:2463–2469.

Maruya, K. A., W. Vetter, S. G. Wakeham, R. F. Lee, and L. Francendese. 2000. Selective persistence and bioaccumulation of toxaphene in a coastal wetland, pp. 164–175. In, Lipnick, R. L., J. L. M. Hermens, K. C. Jones, and D. C. G. Muir (eds.), Persistent, Bioaccumulative, and Toxic Chemicals. American Chemical Society, Washington, DC.

McCormack, M. L. 1994. Reductions in herbicide use for forest vegetation management. *Weed Technology* 8:344–349.

Miles, C. J., G. Leong, and S. Dollar. 1992. Pesticides in marine sediments associated with golf course runoff. *Bulletin of Environmental Contamination and Toxicology* 49:179–185.

Neary, D. G., P. B. Bush, and J. L. Michael. 1993. Fate, dissipation and environmental effects of pesticides in southern forests—a review of a decade of research progress. *Journal of Environmental Toxicology and Chemistry* 12:411–428.

Pait, A. S., D. R. Farrow, J. A. Lowe, and P. A. Pacheco. 1989. Agricultural pesticide use in estuarine drainage areas: A preliminary summary for selected pesticides. Ocean Assessments Division NOS/NOAA, National Oceanic and Atmospheric Administration, Rockville, MD. 134 p.

Pait, A. S., A. DeSouza, and D. Farrow. 1992. Agricultural pesticides in coastal areas: A national summary. Strategic Environmental Assessments Division, ORCA/NOS/NOAA, National Oceanic and Atmospheric Administration, Rockville, MD. 112 p.

Palekar, L. D., K. A. Maruya, J. E. Kostka, and J. Wiegel. 2003. Dehalogenation of 2,6-dibromobiphenyl and 2,3,4,5,6-pentachlorobiphenyl in contaminated estuarine sediment. *Chemosphere* 53:593–600.

Perry, R. and A. E. McIntyre. 1987. Oil and polynuclear aromatic hydrocarbon contamination of road runoff—a comparison of treatment procedures, pp. 474–484. In, Vandermeulen, J. H. and S. E. Hrudey (eds.), Oil in Freshwater: Chemistry, Biology, Countermeasure Technology. Pergamon, New York, NY. 512 p.

Racke, K. D. 1993. Urban pest control scenarios and chemicals, 2–9. In, Racke, K. D. and A. R. Leslie (eds.), Pesticides in Urban Environments. American Chemical Society, Washington, DC.

Ross, P., G. I. Scott, M. H. Fulton, and E. D. Strozier. 1996. Immunoassays for rapid, inexpensive monitoring of agricultural chemicals, pp. 345–367. In, Richardson, M. (ed.,) Environmental Xenobiotics. Taylor and Francis, London.

Ryals, S. C., M. B. Genter, and R .B. Leidy. 1998. Assessment of surface water quality on three eastern North Carolina golf courses. *Journal of Environmental Toxicology and Chemistry* 17:1934–1942.

Sadlon, N. P. 1993. Managing for the birds and more! Is golfing green? The impact of golf courses on the coastal environment. North Carolina Cooperative Extension Service, University of North Carolina, Wilmington, NC. 9 p.

Sanger, D. M., A. F. Holland, and G. I. Scott. 1999a. Tidal creek and salt marsh sediments in South Carolina coastal estuaries: I. Distribution of trace metals. *Archives of Environmental Contamination and Toxicology* 37:458–471.

Sanger, D. M., A. F. Holland, and G. I. Scott. 1999b. Tidal creek and salt marsh sediments in South Carolina coastal estuaries: II. Distribution of organic contaminants. *Archives of Environmental Contamination and Toxicology* 37:445–457.

Scott, G. I., D. S. Baughman, A. H. Trim, and J. Dee. 1987. Lethal and sublethal effects of insecticides commonly found in nonpoint source agricultural runoff to estuarine fish and shellfish, pp. 251–273. In, Vernberg, W. B., F. Thurberg, A. Calabrese, and F. Vernberg (eds.),

Pollution Physiology of Estuarine Organisms. University of South Carolina Press, Columbia.

Scott, G. I., M. H. Fulton, M. Crosby, P. B. Key, J. W. Daugomach, J. Waldren, E. Strozier, C. Louden, G. T. Chandler, T. Bidleman, K. Jackson, T. Hampton, T. Huffman, A. Schulz, and M. Bradford. 1992. Agricultural insecticide runoff effects on estuarine organisms: correlating laboratory and field toxicity test, ecophysiology bioassays and ecotoxicological biomonitoring (Final Report). U.S. Environmental Protection Agency, Gulf Breeze Environmental Research Laboratory, Gulf Breeze, FL.

Selck, H., A. Palmqvist, and V. E. Forbes. 2003. Biotransformation of dissolved and sediment-bound fluoranthene in the polychaete, *Capitella* sp. I. *Environmental Toxicology and Chemistry* 22:2364–2374.

Soma, Y. and M. Soma. 1989. Chemical reactions of organic compounds on clay surfaces. *Environmental Health Perspectives* 83:205–214.

South, D. B. and R. J. Mitchell. 1999. Determining the optimum slash pine seedling size for use with four levels of vegetation management on a flatwoods site in Georgia, USA. *Canadian Journal of Forestry Research* 29:1039–1046.

Southwick, L. M., G. H. Willis, T. E. Reagan, and L. M. Rodriguez. 1995. Residues in runoff and on leaves of azinphosmethyl and esfenvalerate applied to sugarcane. *Environmental Entomology* 24:1013–1017.

Spies, R. B., B. D. Andresen, and D. W. Rice. 1987. Benzthiazoles in estuarine sediments as indicators of street runoff. Nature 327: 697–699.

Stephensen, E., M. Adolfsson-Erici, M. Celander, M. Hulander, J. Pakkonen, T. Hegelund, J. Sturve, L. Hasselberg, M. Bengtsson, and L. Förlin. 2003. Biomarker responses and chemical analyses in fish indicate leakage of polycyclic aromatic hydrocarbons and other compounds from car tire rubber. *Environmental Toxicology and Chemistry* 22:2926–2931.

Sullivan, B. 1997. Early data on the Georgia tidewater. Lower Altamaha Historical Society, Darien, GA. 11 p.

Syngenta. 2003. http:// www.syngentacropprotection-us.com/enviro/ futuretopics/Bravo%20July17.pdf.

Tingle, C. C., J. A. Rother, C. F. Dewhurst, S. Aluer, and W. J. King. 2003. Fipronil: Environmental fate, ecotoxicology, and human health concerns. *Reviews of Environmental Contamination and Toxicology* 176:1–66.

Torstensson, L., E. Borjesson, and B. Arvidsson. 1999. Treatment of bare root spruce seedlings with permethrin again pine weevil before lifting. *Scandinavian Journal of Forestry Research* 14:408–415.

Trim, A. H. and J. M. Marcus. 1990. Integration of long-term fish kill data with ambient water quality monitoring data and application to water quality management. *Environmental Management* 14:389–396.

U.S. Environmental Protection Agency (U.S. EPA). 2003. Persistent, bioaccumulative and toxic (PBT) chemical program. 12 p. http://www.epa.gov.pbt/.

Verschueren, K. 1983. Handbook of environmental data on organic chemicals. 2nd ed. Van Nostrand Reinhold, New York, NY.

Walker, A. A., C. R. Cripe, P. H. Pritchard, and A. W. Bourquin. 1988. Biological and abiotic degradation of xenobiotic compounds in in vitro estuarine water and sediment/water systems. *Chemosphere* 17:2255–2270.

Walker, W. J. and B. Branham. 1992. Environmental impacts of turfgrass fertilization, pp. 105–219. In, Balogh, J. C. and W. J. Walker (eds.), Golf Course Management and Construction — Environmental Issues. Lewis, Baca Raton, FL.

Willis, G. H. and L. L. McDowell. 1982. Review: Pesticides in agricultural runoff and their effects on downstream water quality. *Environmental Toxicology and Chemistry* 1:267–270.

Windom, H., W. Gardner, J. Stephens, and F. Taylor. 1976. The role of methylmercury production in the transfer of mercury in a salt marsh ecosystem. *Estuarine, Coastal and Marine Science* 4:579–583.

Wirth, E. F., P. L. Pennington, J. C. Lawton, M. E. DeLorenzo, D. Bearden, B. Shaddrix, S. Silvertsen, and M. H. Fulton. 2004. The effects of the contemporary-use insecticide (fipronil) in an estuarine mesocosm. *Environmental Pollution* 131:365–371.

Wolfe, D. A. 1992. Selection of bioindicators of pollution for marine monitoring programmes. *Chemical Ecology* 6:149–167.

Zepp, R. G., P. F. Schlotzhauer, and R. M. Sink. 1985. Photosensitized transformation involving electronic energy transfer in natural waters: role of humic substances. *Environmental Science and Technology* 19:74–81.

Models of Coastal Stress: Review and Future Challenges

Thomas C. Siewicki

Summary by Thomas C. Siewicki

Models help explain complex relationships. A current trend is to link models that explain different environmental processes. Hydrodynamic models help describe circulation patterns, tidal flows, and freshwater retention. They are being coupled to pollution runoff models. Sedimentation models predict sources, transport, deposition, and resuspension of particulates. These processes often dictate the fates of adhering chemical contaminants. Nutrient runoff models are being coupled with hydrodynamic models at subwatershed levels. These models frequently incorporate novel surrogates of environmental parameters that are difficult to measure. The current trend for chemical contaminant modeling is to link chemical-process, hydrodynamic, salinity, and sedimentation models to predict sources, transport, fate, and toxicity. Land use characteristics are increasingly included, particularly in urban settings. Geographic information processing will become increasingly important in estuarine modeling of all types. The linking of stressor models with biomass productivity is just beginning. Finally, development of models specific to marine and estuarine conditions lags behind model development for freshwater systems. However, research that links hydrodynamics, sedimentation, nutrient and chemical contamination, geographic information systems (GIS), and other land use techniques to predict effects on saltwater systems is progressing.

10.1 Introduction

There have been many models created to simulate processes that affect or occur within tidal systems. These models were created for different purposes and different settings. It is not possible to identify specific models for given circumstances without understanding what is available and where limitations have been identified. This chapter describes current trends in modeling. It briefly describes examples of models that characterize stresses on estuaries similar to those found in the southeastern United States.

Models allow theory to be developed that better explains processes at work. Historically, the complex and dynamic nature of coastal and estuarine areas has inhibited the ability to model such areas. Fortunately, the advancement of computer and information technologies is allowing researchers to incorporate multi-disciplinary scientific expertise and data into newly developed integrated ecosystem models. This is a review of models used to predict sources, fates, and effects of stressors in estuaries of the southeastern United States, with an emphasis on chemical stressors. This review focuses on models either already used or thought applicable to estuaries in the geographic area approximately bounded by Cape Fear, North Carolina, and Cape Canaveral, Florida, and typified by vast *Spartina* marshes that contain creeks, channels, and barrier islands. Brief descriptions of models and their development are provided for several priority applications. Both riverine and nonriverine estuaries, impacted by a wide range of freshwater flows, are found in this region. Thus, models are discussed that are applicable to a variety of conditions.

Thomann (1998) described three historical stages in model development. The first was from 1925 to 1980, when stressor sources were external to the models with only point sources linked directly. In the second phase, from 1980 to 1990, sediment models were coupled to water column and hydrodynamic/watershed models. In the current, third phase, air sheds and other watershed aspects (e.g., land use) are being incorporated. Future improvements will come from better understanding of ecosystems and from developing consensus between land and estuary use managers (Thomann 1998).

Several reviews on different aspects of environmental modeling have been done. Jorgensen et al. (1995) provided a description of environmental and ecological models. A general review by Chang (1992) described the variety of aquatic models that were available for river basins, estuaries, lakes, reservoirs, nonpoint source pollution storm water management, groundwater, potable water, and wastewater. A more specific review of estuarine flow and water quality models was done by Najarian and Harleman (1989). A review of modeling of nonpoint source pollution in the vadose zone using geographic information systems (GIS) is available (Corwin et al. 1997). The U.S. Environmental Protection Agency lists linear regression models that used urban watershed data (U.S. EPA 1991, 1992). An excellent overview of general modeling considerations was provided by Lung (1989). Some of the most significant stressors identified for estuarine modeling included: salinity changes, suspended solids, microbial agents, dissolved oxygen, biological oxygen demand (BOD), chemical oxygen demand (COD), nutrients, and toxic chemicals (both acute and chronic). Both hydrodynamic and water quality models were included.

Smith (1992) prepared a report on the use of computer models of estuaries by eastern coastal states that identified the models that were most often used and explained why they were popular. State agency modelers have different concerns and needs than federal, private, or academic modelers. These agencies often used models that are based on the U.S. EPA Water Quality Analysis Simulation Program (WASP4) or Enhanced Stream Water Quality Model (QUAL2E) because they often receive U.S. EPA funding. Models used in the southeastern United States included: WASP4, CE-QUAL-W2, HAR-03, Georgia Estuary Model, MIT Transient Water Quality Network, QUAL-2E, Chesapeake Bay Water Quality Model (CBWQ), Hydrological Simulation Program (HSPF), SPAM, AWEST, DEM-DYNDEL, TOXIWASP, Transient Salinity Intrusion Model, Cornell Mixing Zone Expert System (CORMIX), NCWQAP, BLTM, CWQM, the EPA Simplified Math Model, AUTOSS, HEM, JRWQM, SIM, tidal prism (TPM), VMP, PEM, and TAM (Smith 1992). (The reader is reminded that, perhaps more so in modeling than in other areas, acronyms are more recognizable than full names.)

Appropriate model selection should depend upon the objectives of the study, the level of detail required, and model criteria (DeCoursey 1985). Models help identify areas where science is lacking (particularly regarding coupling of processes), facilitate sensitivity analysis of important processes, enable examination of dominant scales of processes, and test water resource management alternatives for implementation (Najarian and Harleman 1989).

10.2 Hydrodynamics

Hydrodynamic models describe circulation patterns and tidal flows in estuaries that influence transport of stressors. For example, the TRIM-3D, three-dimensional model simulated shallow water flow and tidal circulation in estuaries (Casulli and Bertolazzi 1993). The TABS-2 model developed by the U.S. Army Corps of Engineers Waterways Experiment Station characterized shallow water flow in estuaries (Jones and Richards 1992). Lowery (1998) described a model used in Gulf of Mexico estuaries to characterize freshwater retention and flushing over tidal cycles. Wang and Jin (2001) described a GIS model that predicted flooding by using the U.S. Department of Agriculture Soil Conservation Service curve number approach along with flow direction and accumulation. This model might be adaptable to tidal systems. It was similar to the approach used by a consortium of researchers building a geodatabase for hydrologic modeling in a GIS framework (Maidment 2001). Determination of whether to use a circulation or a tidal flow model depends upon the objectives of the study and hydrography of the study area.

One innovative study coupled a pollution runoff model (the Agricultural Non-Point Source [AGNPS]) with a hydrodynamic model (the Simulation of Water Resources for Rural Basin [SWRRB]), and applied them to Charleston Harbor, South Carolina, in an attempt to characterize nutrient loading in a southeastern estuary. This study used GIS to integrate detailed spatial data, particularly about land use, with both models to predict nitrogen loads (Choi 1996). Another group coupled GIS with a discharge model to estimate the contribution of groundwater into shallow coastal bays (Steck et al. 2001). Bhaduri et al. (2000) observed that existing models did not always adequately simulate land use conditions along the coast. They demonstrated that most nonpoint source contaminants accumulate from frequent, small, low-intensity storms rather than infrequent high-intensity storms. Using the Long-Term Hydrologic Impact Assessment (L-THIA) model, they were able to assimilate land use change, climate records, and soils data within a GIS to associate land use change with changes in runoff volume, as well as amounts of inorganic contaminant loading. The U.S. EPA provided a hydrologic model for storm water management called the Storm Water Management Model (SWMM) that estimated both quantity and quality of urban runoff that might apply to estuarine settings. The WASP model has been used, alone or linked to other models, to estimate transport and fate of constituents in water, including BOD, dissolved oxygen, nutrients, eutrophication, and bacterial contamination, as well as both organic and inorganic chemical contamination. Cheng and Smith (1993) provided a review of improvements in estuarine hydrodynamic modeling. They emphasized formulation, numerical methods, spatial and temporal resolution, and computational efficiency. The future of hydrodynamic modeling will couple more model types with greater diversity of data sets.

10.3 Sediment

Several models were used to predict sediment sources, transport, and fate. The CREAMS sediment model was used to determine both sediment loads and degradation (Williams and Nicks 1981). The U.S. EPA Total Maximum Daily Load Universal Soil Loss Equation (TMDL USLE) model also estimated soil loading in a system. Evans and Seamon (1997) used another universal soil loss equation model (RUSLE) coupled with GIS to estimate sediment delivery, storage, and routing in a small rural watershed. The improvements in length-slope factors from the GIS improved the model compared to simulations without the GIS component. Benninger and Wells (1993) used potassium/aluminum ratios and zinc (Zn) and copper (Cu) levels in sediment to determine the source of sediment. Applying the technique to a southeastern estuary, they discovered that fine sediment in the Neuse River (North Carolina) came from Pamlico Sound. Brunk (1997) described turbulent coagulation of sediment. He used phenanthrene, kaolin, and bacterial polymers as estuarine surrogate components (Brunk 1997). Some models simulate vertical particle movement. The rate of sediment resuspension and deposition appeared to control particle association and removal from water (Baskaran and Santschi 1993). Gessler et al. (1999) determined that a three-dimensional model was needed to predict sediment erosion and deposition when significant secondary currents existed due to river bends, crossings, or diversions. The trend is to incorporate sediment models with other models to predict transport and fate of associated stressors. The Three-Dimensional Numerical Model of Hydrodynamics and Sediment Transport in Lakes and Estuaries (SED3D), distributed by the U.S. EPA, modeled both flow and sediment transport. Lung (1989) provided a good overview of important considerations for sediment modeling.

Sediment modeling may be useful for estimating transport and fate of microbiological agents. Although not a focus of this paper, these models have been used to predict indicator microorganisms such as fecal coliform bacteria, enterococcus bacteria, and pathogens of both humans and aquatic organisms when particles of similar size and density were selected. Similarly, some of these models might be useful for simulating dispersion and fate of bioterrorism agents, such as anthrax, smallpox, brucella, rickettsia, and plague. Once these agents enter estuarine systems, they can influence usability of estuaries for fish and shellfish harvests, aquaculture, and recreation. Models can be valuable for preparing emergency responses, focusing resources and directing clean-up operations.

10.4 Nutrients

Numerous regression-based models of nutrient transport and fate, as well as eutrophication, have been developed with varying applicability to the coastal Southeast. An empirical model was derived for the Potomac River to calculate area loadings of pollution from both urban and agricultural areas (Smullen et al. 1978). Runoff samples were analyzed for total nitrogen, total and dissolved kjeldahl nitrogen, nitrite, nitrate, total and dissolved phosphorus, ortho-phosphorus, COD, and both biological and chemical contaminants. The authors used simple linear regression models based on rainfall, impervious surface area, and other land uses. Driver and Tasker (1988) developed regressions for nitrogen and phosphorus constituents in runoff that depend upon rainfall amount, total drainage area, and combinations of impervious surface area, commercial land use, residential land use, nonurban land use, and temperature,. More recently, the Little River (undeveloped river) and Webhannet River (extensively developed) in Maine were compared (Holden 1997). In this study, urban land use was estimated to produce nitrogen and phosphorus loadings 1.4 and 2.5 times higher, respectively, in the developed compared to the undeveloped estuary. The model was more accurate for the undeveloped estuary but predicted 39–83 percent of observed nitrogen in the developed estuary. Developed land drained by small streams and groundwater accounted for much of the difference. Currently available empirical models link land uses to nutrient loading, but few examples are specific to the coasts of the southeastern United States.

Criteria and methods for nutrient modeling that are useful to coastal planners have been developed. A simulation model by Hopkinson and Vallino (1993) explored effects of anthropogenic impacts in watersheds upon spatial patterns of production and respiration in a generalized estuarine system. Effects of variations in ratios of inorganic to organic nitrogen loading, residence time of water in the estuary, degradability of allochthonous organic matter, and ratios of dissolved to particulate organic matter inputs were incorporated. They showed total organic carbon levels in some U.S. rivers have increased three- to five-fold, largely due to channelization and dams. The Watershed Ecosystem Nutrient Dynamics (WEND) model estimated phosphorus export over long time periods and covered entire watersheds (Cassell et al. 2001). Shen (1996) developed a two-dimensional, real-time, integrated hydrodynamic-eutrophication model. The hydrodynamic model provided input to the eutrophication model. A model of the Chesapeake Bay tested pollution-reduction strategies by coupling a water quality model to a 3-D hydrodynamic model, thus coupling a water column model to sediment oxygen demand and a nutrient flux model on an intertidal scale (Cerco et al. 1995). A GIS-based model coupled with an unsaturated root zone transport model was used to assess groundwater transport of nitrate (Lassere et al. 1999). In addition, three submodels were coupled to evaluate the effects of vegetated buffer strips on nutrient transport (Srivastava et al. 1998). Researchers are increasingly using models to characterize the linkages between land use change, nutrient loading, and resulting impacts on receiving water bodies.

Some researchers have developed surrogates for hard-to-measure parameters in nutrient modeling. Lowery (1996) developed methods for estimating subwatershed nitrogen loading surrogates based upon human population, flushing estimates, and eutrophication modeling by multinomial logistic regression. Similarly, Costanza (1996) conducted sub-basin eutrophication modeling of nitrogen loading, flushing rates, and eutrophication models for estuaries and human populations. Sub-basin estimates were more predictive than total estuary estimates (Costanza 1996). Surrogate predictor variables can be less expensive and easier to obtain than direct measurements. Use of surrogates will expand the application of models.

Widely used models have been improved and made more applicable to a broader range of circumstances. The QUAL2E (and QUAL2EU that includes uncertainty analysis) model has been used to analyze dissolved oxygen in streams and rivers. The estuarine new math model, DSM2-QUAL, was developed to incorporate new routines for decay, growth, and interactions among water quality parameters (primary production, dissolved oxygen, phytoplankton, nutrients, and temperature) in the original QUAL model (Rajbhandari 1995). The Pigeon River Allocation Model (PRAM) was run in concert with the QUAL2E model to represent carbon, oxygen, and nutrient dynamics in rivers (Brown and Barnwell 1987). The AGNPS model simulated generation, transport, and deposition of water, sediment, and nutrients in receiving waters due to farm management practices (Young et al. 1987). The GLEAMS model was coupled with a GIS to test for agricultural nitrate pollution. Specific land uses, e.g., vegetable crops and groundwater-irrigated land, increased nitrate contamination in the groundwater (De Paz and Ramos 2002). A GIS-based hydrologic model incorporated riparian land use/land cover variables to explain more of the variation of nutrient export in streams (Baker et al. 2001). Yetter (2001) used the BASINS model output as input into the MOD-FLOW groundwater model to demonstrate nitrogen source loading. The model estimated that most of the nitrogen that affected the estuary came from nitrates in baseflow groundwater, whereas most of the nitrogen in storm water runoff came from atmospheric deposition. Shepherd et al. (1999) evaluated 14 models for estimating phosphorus loss from a catchment and determined that the SWAT model was most suitable. These popular models will continue to be used widely to characterize nutrient loading. Unfortunately, few have been calibrated to southeastern coastal settings.

10.5 Toxic Chemicals

Along with an increasing interest in managing toxic chemical pollution there has been a corresponding increase in chemical modeling. Most models of chemical contaminants in estuaries have examined the distributions and fates of toxic chemicals and most were biogeochemical in nature, i.e., they predict mass transport. Chemical ecotoxicological models have been food chain or trophic level based or mass flow based. Some models that predicted the fates of chemicals included risk assessment components. Many incorporated measures of sedimentation. When contaminants were sorbed to sediment, the sediment has been treated as either a sink, or a source when resuspended into the water column.

Hwang (1995) linked models to produce a hydrodynamic-sediment transport contaminant model. The hydrodynamic model generated velocity measures; the sediment model predicted movements and concentrations of suspended sediment in estuaries and coastal waters; and the toxicity model quantified interactions between sediments and chemicals and impacts on biota (adsorption, desorption, exchange). The U.S. EPA Hydrodynamic, Sediment, and Contaminant Transport (HSCTM2D) Model was a vertically integrated, two-dimensional approach to also link water flow, sediment flow, and contaminant transport. Nichols (1990) modeled the fate of Kepone™ in the James River, Virginia, through fine-sediment dispersal to sediment sinks. He estimated that 42–90 percent of all Kepone™ input remained in this system by entrapment in estuarine circulation and seasonal refluxing, with its ultimate fate being the mid-estuary bed sediment (Nichols, 1990). The Simplified Method Program—Variable Complexity Stream Toxics Model (SMPTOX3) estimated water column and stream bed toxic chemical concentrations as well as waste load allocation estimates. Models that couple hydrodynamic and sediment transport processes will be useful for stressors that sorb to sediment or are particulate in nature

A limited number of environmental processes affect metal disposition. A two-dimensional model of copper distribution and fate illustrated that important processes include advection, dispersion, partitioning with suspended particles, settling, and resuspension of copper (Chen et al. 1996). Cuthbert and Kalff (1993) found that aluminum (Al), iron (Fe), manganese (Mn), Zn, and Cu distributions were related to suspended particulate levels, turbidity, color, temperature, and hydrology. These variables explained large portions of the system variation better for some metals than others. Independent variables included average areal runoff, average suspended particulate matter for most metals, and conductivity and water color for dissolved Fe (Cuthbert and Kalff 1993). Wood (1993) developed a spatially and temporally explicit model (elaMET) to describe the fate and transport of nonconservative metals in estuaries that incorporated advection, dispersion, and transformation. It was used to characterize adsorption kinetics and distribution of Cu, Cd, and Zn (Wood 1993). A two-dimensional model of fate and transport of toxic heavy metals associated with cohesive sediment was also developed. Regnier and Wollast (1993) examined the distribution of nickel (Ni), cobalt (Co), chromium (Cr), Zn, Cu, cadmium (Cd), lead (Pb), and Mn in suspended matter and sediment. Trace metals normalized to Al allowed characterization of the origin of the solids and evaluation of their degree of contamination. Large portions of the metal carried by the estuary were bound to suspended matter, mainly in fine grains, and accumulated in areas of low salinity (Regnier and Wollast 1993). De Smedt et al. (1998) found it necessary to couple four models—hydrodynamic, salt transport, suspended sediment, and metal transport—to adequately simulate the disposition of metals in an estuary. Both empirical and process-based models, usually coupled with suspended sediment transport, have been used to describe the fate of metals in estuaries.

Salinity can be the dominant process for determining the fate of chemicals (Yeats 1993). Al, Mn, Fe, Co, Ni, Cu, and Zn distributions were described by simple relationships with salinity (Yeats 1993). However, application of the model broke down as the salinity regime became more complex. Salinity-based models will be useful in southeastern estuaries where large ranges of salinity occur within a study area.

Many models were specific to the fate of organic chemicals only. Pesticides must often be treated differently because they are intentionally applied. Unlike other chemical contaminants, the quantity of pesticide introduced to a landscape is often known. The semi-empirical model, SWAT, was developed to predict concentrations of agricultural pesticides moving to surface water based upon hydrologic characteristics, soil type, and amount of water moving to streams from rainfall (Brown and Hollis 1996). The Groundwater Loading Effects of Agricultural Management Systems (GLEAMS) simulation model linked climate, soil, topography, nutrient, and pesticide data to assess land management practices (Leonard et al. 1987). Di Guardo et al. (1994) developed an equilibrium model to predict pesticide concentrations in agricultural runoff based on chemical fugacity. The GCSolar model distributed by the U.S. EPA computed photolysis rates and half-lives of pollutants. Ambrose (1987) linked the chemical transport and fate model, TOXIWASP, to the hydrodynamic model, DYNHYD, to calculate upstream migration of seven organic contaminants: chloroform, 1,2 dichloroethane, 1,2 dichloropropane, dimethoxymethane, methylene, perchloroethylene, and trichloroethylene. Tested in the Delaware estuary, it used published chemical property data. Volatilization was the predominant loss mechanism (Ambrose 1987). The HSPF model combined watershed hydrology and water quality for organic pollutants (primarily pesticides) and nutrients. However, many other lake and river models may be useful for estuaries given appropriate adaptation.

The Food and Gill Exchange of Toxic Substances (FGETS) model estimated whole-body concentrations of nonionic organic chemicals from either food or water. Siewicki (1997) used the Exposure Analysis Modeling System (EXAMS-II) (Burns et al. 1982), first developed for modeling transport, fate, persistence, and exposure of dissolved chemicals, to predict environmental fate and exposure of oysters to particulate-bound fluoranthene. The model suggested that land use and boating had the greatest influence on exposure and can be used to predict human health risks. The current trend is to use process-based models to characterize pesticide loading. These will be coupled with improved land use data as well as groundwater loading for chemicals not sorbed to soils.

Land use characteristics are now being considered in models of chemical transport and fate. Empirical models were derived for calculating area loadings of pollution in the Potomac River from both urban and agricultural areas. Simple linear regression models were based upon rainfall, impervious surface area, and land use (Smullen et al. 1978). Driver and Tasker (1988) developed regressions to predict Cd, Cu, Pb, and Zn concentrations in stormwater runoff. These regressions incorporated total rainfall and total drainage area along with temperature, industrial land use, commercial land use, nonurban land use, residential land use, impervious surface area, or temperature, depending upon the metal. Maslia et al. (1994) described environmental assessment and site remediation studies of spatial and temporal chemical distributions that use GIS, simulation models, and demographic databases to automate exposure assessments. They applied SLAM (Steady Layered Aquifer Model) to simulate groundwater flow and CLAM (Contaminant transport in Layered Aquifer Media) to simulate aquifer contamination, and then used census data and GIS to incorporate spatial distributions of human population (Maslia et al. 1994). Porter et al. (1996) used simple linear regression equations to predict oyster tissue fluoranthene concentrations in Murrells Inlet, South Carolina, based on landscape features and surrogate pollution measures. GIS, data management tools, gridding software input and output interfaces, and process models were utilized to predict oil transport, fate, dispersion and bottom deposition, sewage discharges, and water quality impacts from combined sewer overflows (Spaulding and Howlett 1995). A land use-based exposure assessment model was developed for the St. Johns River, Florida

(Siewicki et al. 2003). It was used to estimate risks to resident fauna from current-use pesticides based on land uses. A similar approach could be applied to other areas.

Thomann et al. (1991) modeled fate and bioaccumulation of PCB homologues in striped bass. They found that of the total PCBs discharged into the Hudson River, New York, since 1947, 66 percent was volatilized, 6 percent was stored into sediment, and only a small portion was taken up by fish. More than 90 percent of fish-tissue PCBs resulted from food chain transfer. Rowan and Rasmussen (1992) reviewed empirical models of contaminants in the Great Lakes. They found that researchers were using log-linear multiple regressions to link tissue concentrations to water and sediment characteristics as well as basin-specific ecological attributes. Important factors for determining tissue concentrations of PCBs and DDT were tissue lipid, trophic level, and trophic structure of the food chain, explaining 59 (DDT) to 72 (PCBs) percent of the variability of 25 species (Rowan and Rasmussen 1992). The Oyster Bioaccumulation Model used oyster uptake and growth parameters to estimate radionuclide uptake by oysters near nuclear power plants (Rose 1989). ECOFATE (Gobas 1992) estimated the amount of organic chemical in water, sediment, and aquatic organism tissue.

Toxicity has also been modeled. Combined toxic effects of chemical mixtures on microorganisms were predicted using the Quantitative Structure-Activity Relationship (QSAR) (Nirmalakhandan et al. 1994). QSAR was used to predict concentrations that cause 50 percent growth inhibition, and the investigators found that predicted values agreed with measured values with an $R^2 = 0.8$ (Nirmalakhandan et al. 1994). The WASP4 model was used to predict growth inhibition in fathead minnows (Fikslin 1994). Sensitivity analyses indicated that the model was most sensitive to loading from point sources, model boundaries, and decay rates. It was least sensitive to freshwater inflows, tidal phase, and dispersion coefficients. Copper and Zn interactions seemed to account for toxicity (Fikslin 1994). Another model based on the WASP model was the Partitioning, Mass Balance, and Bioaccumulation Model for Hydrophobic Organic Chemicals in Lake Ontario (Endicott and Cook 1994). This model was used to predict organic chemical concentrations in water, sediment, and biota based upon hydrology, particle transport, chemical property, physicochemistry, toxicity, and bioenergetic data. The U.S. EPA PLUMES and Visual PLUMES models simulate chemical dispersion in water jets and plumes in water and might be particularly applicable in emergencies such as terrorism events. Along with the QSAR and WASP models, distributed toxicity models must be calibrated to southeastern estuarine settings before they will be widely used in the region.

10.6 Population Dynamics

Models of biomass productivity, distribution, and impacts on ecosystems will be coupled to stressor source, transport, fate, and toxicity models in the future. A few examples of biomass models are provided. The Estuarine Phytoplankton Model (Cloern 1991) was used to assess phytoplankton blooms in estuaries based largely upon zooplankton and macrofauna grazing, sinking, and turbulence data. The Dynamic System Model of Plankton Growth and Nutrient Uptake (Kumar 1991) predicted phytoplankton growth by classes of organisms using detailed plankton-nutrient interactions in seawater. Madden and Kemp (1996) described a dynamic simulation model that was developed to elucidate mechanisms responsible for decline of submerged aquatic plants in Chesapeake Bay. Their simulations investigated the influence of phytoplankton and epiphytes on underwater light, balance of limiting resources on growth and productivity, and conditions needed for restoration (Madden and Kemp 1996). Jensen et al. (1998) used nonintrusive remote sensing and digital image processing techniques to model aboveground biomass for entire estuaries in South Carolina. When compared to in situ biomass measurements, near-infrared and middle-infrared bands as well as several vegetation indices were highly correlated.

Modeling of populations of free-swimming fish in open systems is particularly difficult. Kiefer and Do Chi (1996) described a pelagic fish population model using satellite ocean color and thermal imagery, scientific surveys, and fish distribution data. Li (1994) described the differential dynamic programming (DDP) method used to determine optimum freshwater inflows into bays and estuaries to maximize fishery productivity. Another population dynamics model was developed for striped bass to improve predictions of recruitment and other population parameters (Rose and Cowan 1993). Many of the population dynamics models, particularly those incorporating remotely sensed data, will eventually be coupled with exposure and toxicity models to assess impacts of stressors on communities and ecosystems.

Dabrowski (1989) developed a bioenergetic model to simulate young fish growth and survival. The model was based upon density, prey type, temperature, and feeding duration, and is applicable to lakes and marine systems. Spatially explicit bioenergetic models have been developed using in situ measurements of water temperature and prey density to predict fish growth (Luo and Brandt 1993). The coupling of bioenergetic and stressor models to predict ecosystem effects of both vegetation and fish populations, as well as effects on stressor levels and impacts, is just evolving. Estuarine population dynamics models will rely increasingly upon remotely sensed data because of the vast spatial extent and extreme temporal nature of plankton and fish occurrence.

10.7 Additional Considerations

A GIS can incorporate a multiplicity of model components. Advantages of incorporating GIS into nonpoint source pollution modeling include: (1) generation of new information cheaply and quickly, (2) contribution of input data for other models, (3) documentation of data sources and methods, (4) improved diagnostics and error detection of models, and (5) mapped outputs of models. The disadvantages include the difficulty of transposing between different scales and the complexity of spatial statistical analyses (Corwin et al. 1999). GIS coupled with remote sensing is also used in water quality modeling. Regression models were developed using landscape predictors and physiographic variables to predict ($R^2 = 0.75$) water quality (Johnson et al. 2001). The IDOR^{2D} GIS model is a two-dimensional hydrodynamic and pollutant transport model built within a GIS (Tsanis and Boyle 2001) and used in coastal and lake systems. Other researchers have demonstrated the utility of developing GIS-based assessment models for addressing coastal zone issues (Porter et al. 1996; Siewicki 1996; Corbett et al. 1997; Porter et al. 1997). Garbrecht et al. (2001) provided a good overview of considerations in data and model selection for GIS watershed modeling.

Urban or suburban growth patterns will influence future contamination problems as well as the selection of models for their simulation. Weng (2001) demonstrated increased runoff from increased impervious surface coverage using a combination of GIS and remote sensing modeling. Within the southeastern United States, large expanses of silviculture and other agriculture practices along the coast are being replaced with resort style development. These developments are typically sprawling single-home communities along golf courses; they often include service industries. Spray fields are becoming more common, helping to make nonpoint source contamination of estuaries a greater concern than point source contamination. Major contributors of nonpoint source contamination include turf grass management and roadway runoff. The probabilistic "colonization" model of Fagan et al. (2001) allowed for the study of changes in urban landscapes through simulation of housing starts. The model can be applied at different spatial scales. A southeastern coastal development model projected urban growth in the vicinity of Charleston, South Carolina, until 2030 using historical growth patterns from 1973 to 1994 (Allen and Lu 2003). This study combined statistical rule-making and GIS modeling with available information on policy decisions, infrastructure, and population trends (see Kleppel, Becker, Allen and Lu, this volume). The Patuxent Landscape Model was developed to integrate land use changes with a general ecosystem model for use as a policy tool (Binder et al. 2003). Another similar example is the CityPlan model (Gu and Young 1998). The use of landscape change information into land use policy will become more common in the southeastern United States as better assessments of land use impacts on ecosystem health are made.

These models must be applied at different scales to estimate growth impacts adequately. The U.S. EPA and others have supported modeling and decision making at the watershed scale. However, many land use decisions in the southeastern United States are made at the Planned Unit Development (PUD) scale that, cumulatively, may have an even greater overall impact on estuarine health. Models must be developed to address both scales.

The geostatistical technique known as "kriging" and other spatial interpolation techniques were used to develop surface data layers for GIS analysis of parameters that are expensive or time-consuming to measure in large numbers over entire study areas (Porter et al. 1997). Because some measurements will always be made at only a limited number of locations, kriging (in some evolved form), in conjunction with GIS, will likely become even more common for estuarine management in the future. Current research on applications of kriging to estuarine research includes explorations into specialized distance measures specific to estuaries (Little et al. 1997; Rathbun 1998). Improvements in spatial statistical methods integrated in both empirical and process models are needed.

Sensitivity analysis was used to identify model parameters exerting the greatest influence on model results. Hamby (1994) reviewed more than a dozen sensitivity analysis techniques. Likewise, Monte et al. (1996) described principles of Empirically Based Uncertainty

Analysis based upon agreement between models and sets of independent empirical data.

Marine models tend to need more complex descriptions of hydrodynamics than river or lake models. This is due to two-way tidal flows and the open nature of these systems. For the same reasons, most marine models are not yet very sophisticated ecological descriptions. It is likely that ecological considerations of marine models will continue to be a rapidly growing area of research.

10.8 Challenges

The coast of the southeastern United States contains large amounts of undeveloped land due largely to extensive silviculture for the pulp paper industry. Development of this land will not typify past coastal land use patterns. Instead of large, industrialized port cities, the southeastern and Gulf of Mexico coasts are evolving into affluent resort and retirement communities. Instead of steel mills and commercial fishing activities, there are golf courses, shopping malls, and recreational water activities. Instead of urban sewage treatment plants, there will continue to be more spray field discharge of effluents, representing the replacement of point source pollution by nonpoint sources. In addition, more emphasis on problems associated with turf grass management will occur in the form of nutrient and pesticide runoff from shoreline golf courses, lawns, parks, and cemeteries. Models that are selected to characterize estuarine stresses must conform to these changes.

State, county, and local regulatory agencies will use models increasingly to satisfy their own mandates. Models will be used to estimate total maximum daily load (TMDL), urban growth, and antiterrorism planning in addition to the assortment of current uses. A major goal will be rugged, broadly applicable models to test impacts on water from alternative development approaches. These models will primarily address nonpoint source pollution from frequent low-intensity rain events, which contrasts with the approaches often used in the past. These models must be easy to use, generally accepted by a variety of regulatory and other users, and must include training on both their uses and limitations.

Major technical challenges for modelers working at estuarine sites in the southeastern United States are being addressed. The trend toward greater integration of different types of models and data will continue. Even more models will use geodatabase frameworks. Spatial statistical techniques will improve. These methods will be better integrated into GIS-based models. GIS-based models will help document model inputs and allow for easy-to-understand maps of results. Modelers will increasingly rely on surrogate variables that mimic important measurements of the environment and are easy and inexpensive to estimate. These models must depend ever more upon remotely sensed data, particularly when study areas are very large. Models must include improved sensitivity and uncertainty analyses to help identify causes of problems and levels of confidence. Researchers must devise better ways to apply models at different scales, particularly the scales of watersheds and PUDs in the southeastern United States.

Shallow groundwater models will be used even more now that the importance of groundwater contributions of nonsorbed chemical and microbial contaminants to mesotidal estuaries is better understood. Hydrologic and sediment models will be coupled and used to address surface loading of particulate-associated contaminants. Both surface and groundwater models must be better integrated with urban growth models for long-range forecasting and planning. Widely distributed models must be tested in southeastern U.S. semi-tropical, porous soil with high rainfall settings to convince regulators of their usefulness. Finally, there needs to be a movement toward land use-based ecological risk assessment because more policy will be influenced by land use-associated risks.

References

Allen, J. S. and K. S. Lu. 2003. Modeling and prediction of future urban growth in the Charleston region of South Carolina: A GIS-based integrated approach. *Conservation Ecology* 8:1–20

Ambrose, R. B. 1987. Modeling volatile organics in the Delaware Estuary. *Journal of Environmental Engineering* 113:703–721.

Baker, M. E., M. J. Wiley, and P. W. Seelbach. 2001. GIS-based hydrologic modeling of riparian areas: implications for stream quality. *Journal of the American Water Resources Association* 37:1615–1628.

Baskaran, M. and P. H. Santschi. 1993. The role of particles and colloids in the transport of radionuclides in coastal environments of Texas. *Marine Chemistry* 43:95–114.

Benninger, L. K. and J. T. Wells. 1993. Sources of sediment to the Neuse River estuary, North Carolina. *Marine Chemistry* 43:137–156.

Bhaduri, B., J. Harbor, B. Engel, and M. Grove. 2000. Assessing watershed-scale, long-term hydrologic impacts of land-use change using a GIS-NPS model. *Environmental Management* 26:643–658.

Binder, C., R. M. Boumans, and R. Costanza. 2003. Applying the Patuxent Landscape Unit Model to human dominated ecosystems: The case of agriculture. *Ecological Modelling* 159:161–177.

Brown, L. C. and T. O. Barnwell. 1987. The enhanced stream water quality model QUAL2E and QUAL2E-UNCAS. Documentation and user model. U.S. EPA-600–3-87–007. U.S. Environmental Protection Agency. 189 p.

Brown, D. and J. M. Hollis. 1996. SWAT—A semi-empirical model to predict concentrations of pesticides entering surface waters from agricultural land. *Pesticide Science* 47:41–50.

Brunk, B. K. 1997. Turbulent coagulation of particles smaller than the length scales of turbulence and equilibrium sorption of phenanthrene to clay: Implications for pollutant transport in the estuarine water column. Ph.D. dissertation, Cornell University, Ithaca, NY. 379 p.

Burns, L. A., D. M. Cline, and R. R. Lassiter. 1982. Exposure Analysis Modeling Systems (EXAMS) User Manual and System Documentation. EPA 600/3/82/023. U.S. Environmental Protection Agency Environmental Research Laboratory. 443 p.

Cassell, E. A., R. L. Kort, D. W. Meals, S. G. Aschmann, J. M. Dorioz, and D. P. Anderson. 2001. Dynamic phosphorus mass balance modeling of large watersheds: Long-term implications of management strategies. *Water Science and Technology* 43:153–162.

Casulli, V. and E. Bertolazzi. 1993. TRIM_3D: A three-dimensional model for accurate simulation of shallow water flow, pp. 1988–1996. In, Shen, H. W., S. T. Su, and F. Wen. (eds.), Hydraulic Engineering '93. American Society of Civil Engineers, San Francisco, CA.

Cerco, C. F., B. H. Johnson, M. Dortch, and R. Hall. 1995. Environmental modeling of Chesapeake Bay. *Cray Channels* 17:14–17.

Chang, S.-Y. 1992. System analysis. *Water Environment Research* 64:297–301.

Chen, C. W., D. Leva, and A. Olivieri, 1996. Modeling the fate of copper discharged to San Francisco Bay. *Journal of Environmental Engineering* 122:924–934.

Cheng, R. T. and P. E. Smith, 1993. Recent developments in three-dimensional numerical estuarine models, pp. 1982–1987. In, Shen, H. W., S. T. Su, and F. Wen. (eds.), Hydraulic Engineering '93. American Society of Civil Engineers, San Francisco, CA.

Choi, K.-S. 1996. Modeling non-point sources runoff in a coastal area. Ph.D. dissertation. University of South Carolina, Columbia, SC.

Cloern, J. 1991. Tidal stirring and phytoplankton bloom dynamics in a estuary. *Journal of Marine Research* 49:203–221.

Corbett, C., M. Wahl, D. E. Porter, D. Edwards, and C. Moise. 1997. Nonpoint source runoff modeling: A comparison of a forested watershed and an urban watershed on the coast of South Carolina. *Journal of Experimental Marine Biology and Ecology* 213:133–149.

Corwin, D. L., P. J. Vaughan, and K. Loague. 1997. Modeling nonpoint source pollution in the vadose zone with GIS. *Environmental Science and Technology* 31:2157–2175.

Corwin, D. L., K. Loague, and T. R. Ellsworth, 1999. Current and future trends in the development of integrated methodologies for assessing non-point source pollutants. *Geophysical Monograph* 108:343–361.

Costanza, R. 1996. Contributions to estuarine eutrophication modeling: Watershed population estimation methodology, estuarine flushing model, and eutrophication model (nitrogen loading, salinity). Ph.D. dissertation. University of Maryland, College Park, MD.

Cuthbert, I. D. and J. Kalff. 1993. Empirical models for estimating the concentrations and exports of metals in rural rivers and streams. *Water, Air, and Soil Pollution* 71:205–230.

Dabrowski, K. 1989. Formulation of a bioenergetic model for coregorine early life history. *Transactions of the American Fisheries Society* 118:138–150.

DeCoursey, G. D. 1985. Mathematical models for nonpoint water pollution control. *Journal of Soil and Water Conservation* 40:408–413.

De Paz, J. M. and C. Ramos 2002. Linkage of a geographical information system with the Gleams model to assess nitrate leaching in agricultural areas. *Environmental Pollution* 118:249–258.

De Smedt, F., V. Vuksanovid, S. Van Meerbeeck, and D. Reyns. 1998. A time-dependent flow model for heavy metals in the Scheldt estuary. *Hydrobiologia* 366:143–155.

Di Guardo, A., D. Calamari, G. Zanin, A. Consalter, and D. Mackay. 1994. Fugacity model of pesticide runoff to surface water: Development and validation. *Chemosphere* 28:511–531.

Driver, N. E. and G. D. Tasker. 1988. Techniques for estimation of storm-runoff loads, volumes and selected constituent concentrations in urban watersheds in the United States, pp. 88–191. U.S. Geological Survey Open File Report 88–191. U.S. Geological Survey.

Endicott, D. D. and P. M. Cook. 1994. Modeling the partitioning and bioaccumulation of TCDD and other hydrophobic organic chemicals in Lake Ontario. *Chemosphere* 28:75–87.

Evans, J. E. and D. E. Seamon. 1997. A GIS model to calculate sediment yields from a small rural watershed, Old Woman Creek, Erie and Huron Counties, Ohio. *Ohio Journal of Science* 97:44–52.

Fagan, W. F., E. Meir, S. S. Carroll, and J. Wu. 2001. The ecology of urban landscapes: Modeling housing starts as a density-dependent colonization process. *Landscape Ecology* 16:33–39.

Fikslin, T. J. 1994. Modeling toxicity to aquatic life in the tidal Delaware River (*Ceriodaphnia dubia, Pimephales promelas*). Ph.D. dissertation. Rutgers, State University of New Jersey. 254 p.

Garbrecht, J., F. L. Ogden, P. A. DeBarry, and D. R. Maidment. 2001. GIS and distributed watershed models. I: Data coverages and sources. *Journal of Hydraulic Engineering* 6:506–514.

Gessler, D., B. Hall, M. Spasojevic, F. Holly, H. Pourtaheri, and N. Raphelt. 1999. Application of 3D mobile bed, hydrodynamic model. *Journal of Hydraulic Engineering* 7:737–749.

Gobas, F. A. P. C. 1992. Modeling the accumulation and toxic impacts of organic chemicals in aquatic food-chains, pp. 129–153. In, Gobas, F. A. P. C., and J. A. McCorquodale (eds.), Chemical Dynamics in Fresh Water Ecosystems. Lewis, Boca Raton, FL.

Gu, K. and W. Young. 1998. Verifying and validating a land use—transport—environment model. *Transportation Planning and Technology* 21:181–202.

Hamby, D. M. 1994. A review of techniques for parameter sensitivity analysis of environmental models. *Environmental Monitoring and Assessment* 32:135–154.

Holden, W. F. 1997. Fresh water, suspended sediment and nutrient influx to the Little River and Webhannet River estuaries, Wells, Maine. Ph.D. dissertation. Boston University.

Hopkinson, C. S. and J. J. Vallino. 1993. The relationships among man's activities in watersheds and estuaries: A model of runoff effects on patterns of estuarine community. *Estuaries* 18:598–621.

Hwang, B. 1995. Modeling remobilization of sediment-bound contaminants in sediments and their fate and transport in overlying waters. Ph.D. dissertation. University of Virginia, Charlottesville, VA. 175 p.

Jensen, J. R., D. E. Porter, C. Coombs, B. Jones, D. White, and S. Schill. 1998. Extraction of smooth cordgrass (*Spartina alterniflora*) biomass and leaf area index parameters from high resolution imagery. *Geocarto International* 13:25–34.

Johnson, G. D., W. L. Myers, and G. P. Patil. 2001. Predictability of surface water pollution loading in Pennsylvania using watershed-based landscape measurements. *Journal of the American Water Resources Association* 37:821–835.

Jones, N. L. and D. R. Richards. 1992. Mesh generation for estuarine flow modeling. *Journal of Waterway, Port, Coastal and Ocean Engineering* 118:599–614.

Jorgensen, S. E., B. Halling-Sorensen, and S. N. Nielsen. 1995. Handbook of Environmental and Ecological Modeling. Lewis, Boca Raton, FL. 672.

Kiefer, D. A. and T. Do Chi. 1996. An environmental model for small pelagics, pp. 1–18. Workshop on the coastal pelagic resources of the upwelling exosystem of Northwest Africa: Research and Predictions. FAO FI/TCP/MOR/4556(A). Food and Agriculture Organization, Rome, Italy.

Kumar, S. K., W. F. Vincent, P. C. Austin, and G. C. Wake. 1991. Picoplankton and marine food chain dynamics in a variable mixed layer: a reaction-diffusion model. *Ecological Modelling* 57:193–219.

Lassere, F., M. Razack, and F. Poitiers. 1999. GIS-linked model for the assessment of nitrate contamination in groundwater. *Journal of Hydrology* 224:81–90.

Leonard, R. A., W. G. Knisel, and D. A. Still. 1987. GLEAMS: Groundwater Loading Effects of Agricultural Management Systems. *Transactions of the American Society of Agricultural Engineers* 30:1403–1418.

Li, G. 1994. Differential dynamic programming for estuarine management. Ph.D. dissertation. Arizona State University, Tempe, AZ.

Little, L. S., D. Edwards, and D. E. Porter. 1997. Kriging in estuaries: As the crow flies, or as the fish swims? *Journal of Experimental Marine Biology and Ecology* 213:1–11..

Lowery, T. A. 1996. Contributions to estuarine eutrophication modeling: Watershed population estimation methodology, estuarine flushing model, and eutrophication model. Ph.D. dissertation. University of Maryland, College Park, MD.

Lowery, T. A. 1998. Difference equation-based estuarine flushing model application to US Gulf of Mexico estuaries. *Journal of Coastal Research* 14:185–195.

Lung, W.-S. 1989. Water Quality Modeling III. Application to Estuaries. CRC Press, Boca Raton, FL. 208 p.

Luo, J. and S. B. Brandt. 1993. Bay anchovy *Anchoa mitchilli* production and consumption in mid-Chesapeake Bay based on a bioenergetics model and acoustic measures of fish abundance. *Marine Ecology Progress Series* 98:223–236.

Madden, C. J. and W. M. Kemp. 1996. Ecosystem model of an estuarine submersed plant community: Calibration and simulation of eutrophication responses. *Estuaries* 19:457-474.

Maidment, D. R. 2001. Arc Hydro: GIS for Water Resources. ESRI Press, Redlands, CA. 220 p.

Maslia, M. L., M. M. Aral, R. C. Williams, A. S. Susten, and J. L. Heitgerd. 1994. Exposure assessment of populations using environmental modeling, demographic analysis and GIS. *Water Resources Bulletin* 30:1025–1041.

Monte, L., L. Haakanson, U. Bergstroem, and J. Brittain. 1996. Uncertainty analysis and validation of environmental models: The empirically based uncertainty analysis. *Ecological Modelling* 91:139–152.

Najarian, T. O. and D. R. F. Harleman. 1989. Role of models in estuarine flow and water quality analysis, pp. 351–373. In, Neilson, B. J., A. Kuo, and J. Brubaker (eds.), Estuarine Circulation. Aquatic Sciences and Fisheries Abstracts. Humana Press, Totowa, NJ.

Nichols, M. M. 1990. Sedimentologic fate and cycling of Kepone in an estuarine system: Example from the James River estuary. *The Science of the Total Environment* 97/98:407–440.

Nirmalakhandan, N., V. Arulgnanendran, M. Mohsin, B. Sun, and F. Cadena. 1994. Toxicity of mixtures of organic chemicals to microorganisms. *Water Resources* 28:543–551.

Porter, D. E., W. K. Michener, T. Siewicki, D. Edwards, and C. Corbett. 1996. Utilizing the tools of geographic information processing to assess the impacts of urbanization on a localized coastal estuary: A multi-disciplinary approach, pp. 355–388. In, Vernberg, F. J., W. B. Vernberg, and T. Siewicki (eds.), Urbanization in Southeastern Estuaries. Belle W. Baruch Library in Marine Science. University of South Carolina Press, Columbia.

Porter, D. E., D. Edwards, G. Scott, B. Jones, and W. S. Street. 1997. Assessing the impacts of anthropogenic and physiographic influences on grass shrimp in localized salt-marsh estuaries. *Aquatic Botany* 58:289–306.

Rajbhandari, H. L. 1995. Dynamic simulation of water quality in surface water systems utilizing a lagrangian reference frame. Ph.D. dissertation. University of California, Davis. 264 p.

Rathbun, S. L. 1998. Spatial modeling in irregularly shaped regions: Kriging estuaries. *Environmetrics* 9:109–130.

Regnier, P. and R. Wollast. 1993. Distribution of trace metals in suspended matter of the Scheldt Estuary. *Marine Chemistry* 43:3–19.

Rose, K. A., R. I. McLean, and J. K. Summers. 1989. Development and monte carlo analysis of an oyster bioaccumulation model applied to biomonitoring data. *Ecological Modelling* 45:111–132.

Rose, K. A. and J. H. Cowan. 1993. Individual-based model of young-of-the-year striped bass population dynamics. I. Model description and baseline simulations. *Transactions of the American Fisheries Society* 122:415–438.

Rowan, D. J. and J. B. Rasmussen. 1992. Why don't Great Lakes fish reflect environmental concentrations of organic contaminants?— An analysis of between-lake variability in the ecological partitioning of PCBs and DDT. *Journal of Great Lakes Research* 18:724–741.

Shen, J. 1996. Water quality modeling as an inverse problem (estuary, eutrophication). Ph.D. dissertation. The College of William and Mary, Williamsburg, VA. 140 p.

Shepherd, B., D. Harper, and A. Millington. 1999. Modelling catchment-scale nutrient transport to watercourses in the U.K. *Hydrobiologia* 395/396:227–238.

Siewicki, T. C. 1997. Environmental modeling and exposure assessment of sediment-associated fluoranthene in a small, urbanized, non-riverine estuary. *Journal of Experimental Marine Biology and Ecology* 213:71–94.

Siewicki, T. C., E. Boyce, and K. Phillips. 2003. Forecasting ecological risks in the St. Johns River, Florida from land uses. U.S. National Oceanic and Atmospheric Administration. 20 p. http://www.chbr.noaa.gov/easi.

Smith, K. 1992. Report on the usage of computer models of estuaries by eastern coastal states. SRS-10 80. North Carolina Water Resources Research Inst., Raleigh, NC.

Smullen, J. T., J. P. Hartigan, and T. J. Grizzard. 1978. Assessment of runoff pollution in coastal watersheds, pp. 840–857. In, Watkinson, R. J. (ed.), Coastal Zone '78. Aquatic Sciences and Fisheries Abstracts, New York.

Spaulding, M. L. and E. Howlett. 1995. A shell based approach to marine environmental modeling. *Journal of Marine Environmental Engineering* 1:175–189.

Srivastava, P., T. A. Costello, D. R. Edwards, and J. A. Ferguson. 1998. Validating a vegetated filter strip performance model. *Transactions of the ASAE* 41:89–95.

Steck, K. P., T. E. McKenna, A. S. Andres, and T. L. DeLiberty. 2001. Use of a GIS model to estimate ground-water discharge to Rehobeth and Indian River bays, Delaware. *Abstracts of Geological Society of America* 33:319.

Thomann, R. V. 1998. The future "golden age" of predictive models for surface water quality and ecosystem management. *Journal of Environmental Engineering—ASCE* 124:94–103.

Thomann, R. V., J. A. Mueller, R. P. Winfield, and C.-R. Huang. 1991. Model of fate and accumulation of PCB homologues in Hudson Estuary. *Journal of Environmental Engineering* 117:161–178.

Tsanis, I. K. and S. Boyle. 2001. A 2D hydrodynamic/pollutant transport GIS model. *Advances in Engineering Software* 32:353–361.

U.S. Environmental Protection Agency (U.S. EPA). 1991. Modeling of non-point source water quality in urban and non-urban areas. EPA 600-3-91-039. U.S. Environmental Protection Agency.

U.S. Environmental Protection Agency (U.S. EPA). 1992. Compendium of watershed-scale models for TMDL development. EPA 841-R-92-002. U.S. Environmental Protection Agency. 96 p.

Wang, X. and J. Jin. 2001. Assessing the impact of urban growth on flooding with as integrated curve number-flow accumulation approach. *International Water* 26:215–222.

Weng, Q. 2001. Modeling urban growth effects on surface runoff with the integration of remote sensing and GIS. *Environmental Management* 28:737–748.

Williams, J. R. and A. D. Nicks. 1981. CREAMS Hydrology Model—Option One. Applied Modeling Catchment Hydrology. In, Singh, V. P. (ed.) Proceedings of the International Symposium on Rainfall-Runnoff Modeling. Mississippi State University, Mississippi State, MS.

Wood, T. M. 1993. Numerical modeling of estuarine geochemistry (trace metals, sediment). Ph.D. dissertation. Oregon Graduate Institute of Science and Technology, Beaverton, OR. 225 p.

Yeats, P. A. 1993. Input of metals to the North Atlantic from two large Canadian estuaries. *Marine Chemistry* 43:201–209.

Yetter, C. 2001. Source identification and modeling the transport of nitrogen into the St. Jones estuary, 2. Proceedings of the 2nd Biennial Coastal GeoTools Conference, January 8–11, 2001. National Oceanic and Atmospheric Administration.

Young, R. A., C. A. Onstad, D. D. Bosch, and W. P. Anderson. 1987. AGNPS, Agricultural nonpoint source pollution model—a water-

Alternatives to Coliform Bacteria as Indicators of Human Impact on Coastal Ecosystems

Marc E. Frischer and Peter G. Verity

Summary by David A. Bryant

The presence of coliform bacteria in coastal waters is evidence of fecal contamination, and fecal contamination is often taken as an indication that pathogens that cause human disease are also present. Many agencies that regulate the use of surface waters use a coliform standard to determine whether the water is safe for drinking or swimming and if it is advisable to eat shellfish or finfish harvested from those waters. However, recently developed scientific techniques offer new indicators for a range of other environmental conditions, in addition to fecal contamination.

Recently, the use of bacteria from the genus *Enterococcus* has begun to supplant the use of coliform bacteria as an indicator of fecal contamination. Both the U.S. EPA and the International Shellfish Safety Committee are considering adopting enterococci as a regulatory indicator, and several studies have demonstrated that these bacteria are better indicators of gastrointestinal illness associated with swimming than are coliform bacteria. Both coliform and enterococci, however, have a significant drawback: they cannot be used to distinguish between fecal contamination that originates with humans and contamination that originates with other animals. It would not be possible to tell, for instance, whether high levels of coliform or enterococci in an estuary are the result of a sewage facility leak or animal fecal runoff from farms. This ambiguity makes it difficult to identify the source of a problem.

Advances in microbiology are beginning to suggest a range of more sophisticated indicators for fecal contamination that resolve some of the problems that coliform and enterococci standards do not, and may offer additional information. A short summary of some of the most promising new proposed indicators follows.

Streptococcus bovis is a fecal bacterium found in warm-blooded animals and birds but is not prevalent in human feces. Another of its characteristics is that it does not survive for very long in fresh or saline water. Therefore, its presence indicates very specifically that contamination is recent and from animal sources.

Bifidobacteria are the third most numerous bacteria in the human intestine, but they are absent or have a low prevalence in animal feces. The presence of these bacteria indicates human fecal contamination. However, to date, most studies of bifidobacteria indicators have been conducted in freshwater environments.

Clostridium perfringens are present in both human and animal feces and thus cannot be used to differentiate between the two. However, the fact that they release spores during reproduction offers an interesting possibility. Because the spores remain viable in sediment for thousands of years, sediment core samples can be used to reconstruct the pollution history of an estuary.

Bacteriophages are viruses that infect bacteria. While it is currently difficult, expensive, and time-consuming to test waters directly for human pathogens, it is relatively easy and inexpensive to test for bacteriophages. The male-specific RNA coliphage (FRNA phage) shares many traits with viruses carried by human feces. Studies suggest that this bacteriophage is a good indicator for human-specific pathogens.

In addition to indicators of human pathogens, scientists are identifying microbial indicators of other aspects of ecosystem health. Two particularly promising areas of research are: (1) finding indicators that would tell regulatory agencies when an estuary is being overloaded with nutrients, a condition called eutrophication; and (2) identifying indicators of certain toxic chemicals. A larger goal is to develop a unified picture of the microbial composition of an estuary so that scientists can assess the full range of effects that human activity has on estuarine environments.

Bacteria play a central role in making waterborne nutrients available to larger organisms in an estuary. At this basic level, bacteria respond to nutrient loading by rapidly increasing their numbers. Studies suggest that long-term monitoring of total bacterial biomass, a relatively simple measurement to make, offers a reliable means of assessing eutrophication. Such monitoring can also be used to evaluate the success of mitigation efforts.

Another way to assess eutrophication is to evaluate the complex mix of species in estuarine microbial communities. The composition of these communities changes as nutrient levels wax and wane. For example, elevated nutrient levels favor smaller species of phytoplankton, some of which can produce blooms that result in fish kills. Cyanobacteria may also thrive in eutrophic conditions. Because of the relatively predictable response of microbial communities to nutrient loading, long-term monitoring can be a reliable indicator of eutrophication in coastal waters.

Luminescent bacteria can account for as much as 10 percent of all bacteria in an estuary. Many chemical contaminants can suppress the numbers of luminescent bacteria. These contaminants include heavy metals, organic solvents, detergents, pesticides, alcohols, straight-chained hydrocarbons, polychlorinated biphenyls, and polycyclic aromatic hydrocarbons. Because such a wide range of contaminants affects luminescence, monitoring luminescent bacteria will not necessarily help identify the source and type of contaminant. It can serve, however, as an index of the general level of chemical contamination in an estuary.

As new techniques such as molecular probes and genetic finger-printing are developed to rapidly identify specific bacteria or to fingerprint the composition and structure of microbial communities, it is becoming increasingly possible to monitor a larger number of specific organisms in the complex microbial mix of the estuary. However, scientists are just beginning to characterize total community structure and to relate that structure to environmental inputs. As work progresses, the small suite of indicators now in use is likely to grow in size and sophistication.

11.1 Introduction—Bacterial Indicators

As cholera and typhoid reached epidemic proportions almost 150 years ago, the fact that water was a vehicle for disease was recognized and proven (Snow 1855). During this period (the late 1800s and early 1900s) the understanding of the nature and causes of waterborne diseases increased rapidly, and significant efforts were made to develop analytical and statistical methods to determine the public health risk of contaminated waters. A primary result of these efforts has been the development of reliable assays based on specific indicator organisms to serve as sentinels of sewage contamination and the development of technology and public works to mitigate these problems. The basic criteria for appropriate water quality indicators are that they should be specifically associated with the contaminant of concern, that they should be easily and reliably detectable, and that they should behave similarly to the pathogen or contaminant of interest. The intent of indicator systems is that they provide information about the type and source of contamination, the risks associated with contamination, and the efficacy of any management efforts implemented to mediate the problem.

As discussed by Fletcher et al. (1998), coliform bacteria have been extensively utilized as a primary bacterial indicator of human sewage contamination and as a surrogate for a number of disease-causing bacteria, protozoans, and viruses (Table 11.1). Coliform bacteria are the current regulatory standard for determining bacterial water quality in shellfish harvesting areas of estuaries throughout the United States, although they have recently been replaced by enterococci for assessment of recreational bathing waters (U.S. EPA 1986). Concentrations of less than 70 total coliform and 14 fecal coliform bacteria per 100 ml of water are considered to be indicative of water quality conditions that insure a high degree of confidence for safe harvesting and consumption of shellfish from estuarine waters (Kantor and Rhodes 1994; U.S. Public Health Service 1995). The advantages of the coliform bacteria as water quality indicators are that simple laboratory procedures for their detection are available, they can be grown easily under common laboratory conditions, they are associated with sewage, and that relatively long-term continuous coliform monitoring data sets exist. However, coliform bacteria can be associated with sources other than humans, and thus the presence of high levels of coliforms may not be an accurate indicator of human pathogens. Additionally, in estuarine waters, coliform bacteria are quickly inactivated and therefore their utility as indicators of pathogens that survive or thrive in saline waters is limited. Coliform bacteria such as *E. coli* typically do not survive longer than 24 hours in saline water, whereas bacterial pathogens such as *Vibrio* or viruses can survive much longer (Kaspar and Tamplin 1993; Mezrioui et al. 1995). Alternatively, in freshwater environments coliform bacteria may actually grow and therefore may be of limited utility for indicating a contamination source (Toranzos 1991; Jamieson et al. 2003; Whitman and Nevers, 2003). Furthermore, sewage contamination represents only one type of contamination that has been associated with human activity, and therefore coliform bacteria are not suitable indicators for other types of human impact on coastal waters.

To evaluate and predict the environmental impact of human activity on coastal systems, it is logical to focus on microorganisms for two reasons. First, the composition and structure of microbial communities are fundamental indicators of ecosystem status, and second, microbial communities are a central ecosystem component, integral to the functioning of all biogeochemical processes. Microorganisms are responsible for the regeneration of nutrients and the transfer of primary production from phytoplankton to microzooplankton and larger organisms (Sherr and Sherr 2000).

Table 11.1. Examples of Waterborne Diseases for Which Coliforms Serve as Indicators.

Pathogen	Type of Organism	Disease
Shigella spp.	Bacteria	Shigellosis
Salmonella typhimurium	Bacteria	Salmonellosis
Salmonella typhi	Bacteria	Typhoid fever
Escherichia coli	Bacteria	Gastroenteritis
Camplyobacter jejuni	Bacteria	Gastroenteritis
Vibrio spp.	Bacteria	Gastroenteritis
Pseudomonas	Bacteria	Opportunistic infections
Legionella	Bacteria	Pontiac fever
Leptospira	Bacteria	Leptospirosis
Hepatitis A	Virus	Hepatitis
Norwalk-like agent	Virus	Gastroenteritis
Rotavirus	Virus	Gastroenteritis
Polioviruses	Virus	Poliomyelitis
Giardia lamblia	Protozoan	Giardiasis
Cryptosporidium	Protozoan	Cryptospordiosis
Entamoeba histolytic	Protozoan	Amebiasis
Ascaris spp.	Helminth	Ascariasis
Ancylostoma spp.	Helminth	Anaemia
Necator sp.	Helminth	Anaemia
Trichuris spp	Helminth	Gastroenteritis

The use of alternative (noncoliform) microbial indicators of environmental impact has been examined in geographically and environmentally diverse systems. In general, alternative microbial indicators have been evaluated as indicators of point and nonpoint sources of fecal pollution (human and nonhuman), eutrophication, soil erosion, and a wide range of different types of chemical contamination. Many alternative indicator organisms have been proposed, and examples of these studies are listed in Table 11.2, with key references

Table 11.2. Alternative Microbial Indicators.

Indicator	Contaminant	Comments	Key Reference
Fecal Pollution			
Clostridium perfringens Monitoring	Fecal Pollution	Long time scale	Edberg et al., 1997
Enterococcus	Fecal Pollution	Persistance greaterthan *E.coli*	Edberg et al., 1997
Streptococcus bovis	Fecal Pollution	Farm Animals	Jagals, 1997
Rhodococcus coprophilus	Animal Fecal Pollution	High Specificity for Animal Source Material	Jagals et al., 1995
Sorbitol- fermenting- bifidobacteria	Human Fecal Pollution	High Specific- ity for Human Source Material	Jagals and Grabow, 1996
Bacteroides fragilis phages	Remote Fecal Pollution (viral indicators)	Persistence model Virus	Lucena et al., 1996
Coliphages	Fecal Pollution	Persistence	Borrego et al., 1990
Staphylococcus aureus	Fecal Pollution	Pathogen	Ashbolt et al., 1993
Campylobacters	Fecal Pollution	Pathogen	Ashbolt et al., 1993
Pseudomonas spp	Fecal Pollution Eutrophication	Human Patho- gen Specific contaminants	Milner and Goulder, 1985 (many others)

(*Continued*)

Table 11.2. Alternative Microbial Indicators. (Continued)

Indicator	Contaminant	Comments	Key Reference
Runoff and Erosion			
Pseudomonas aeruginosa	Urban Runoff	Human Pathogen	Guimaraes et al.,1993
Soil Bacteria	Soil Contamination (erosion)	Indigenous soil specific bacteria	Madsen et al., 1992
(Bacillus mycoides, Myxococcus xanthus Chromobacterium violaceum)			
Eutrophication			
Aeromonas spp.	Eutrophication	Human Pathogen (Not Effective)	Rhodes and Kantor, 1994
Heterotrophic Bacteria	Eutrophication	Ubiquitous	Sabrilli et al., 1997
Cyanobacteria Blooms	Eutrophication	N:P > 20	Paerl and Zehr, 2000
Chemical Pollution			
Pseudomonas putida	Chemical Pollution	Indigenous organism (not effective)	Lemke et al., 1997
Acinetobacter spp.	Chemical Pollution	Indigenous Organism	Lemke et al., 1997
Burkholderia cepacia	Chemical Pollution	Indigenous organism (not effective)	Lemke et al., 1997
Purple Nonsulfur Bacteria	Chemical Pollution (sulfur compounds)	Indigenous organism	Sinha and Banerjee, 1997
Luminescent Bacteria	Chemical Pollution	Indigenous organism	Ramaiah and Chandramohan, 1993

cited. Of these proposed organisms, enterococci have gained general acceptance as indicators of human sewage contamination in drinking and recreational waters (Sinton et al. 1993; WHO 1996; WHO 2001), although numerous concerns have also been voiced (Pinto et al. 1999; Edberg et al. 2000).

11.2 Alternative Microbial Indicators of Fecal Contamination

The presence of enteric indicator organisms does not necessarily indicate human contamination, as livestock and wildlife are also a source. Resolving urban (human) and agricultural (animal) inputs has presented a significant challenge to coliform indicator systems. One strategy that has been pursued to overcome these limitations has been the evaluation of alternate bacterial species and indicator organisms. A number of potential indicator species have been proposed as specific human and nonhuman indicators. For example, organisms such as *Rhodococcus coprophilus* and *Streptococcus bovis* (Oragui and Mara 1983; Mara and Oragui 1985) are thought to be specific for domestic animals, while other organisms, such as sorbitol-fermenting bifidobacteria and *Bacteroiides fragilis* phages are believed to be specific for human fecal pollution (Grabow et al. 1995; Jagals and Grabow 1996; Lucena et al. 1996). It has been demonstrated via several field studies that these indicator species are correlated with contamination associated with particular types of pollution inputs, including storm runoff and inadequate sewerage (e.g., Jagals et al. 1995). Some of these alternative microbial indicators of fecal contamination are discussed below.

11.2.1 Fecal Streptococci and Enterococci

Fecal streptococci are a group of taxonomically diverse streptococci that are Gram-positive, catalase negative, non-spore forming, facultative anaerobes associated with the digestive tract of animals and humans (Carson et al. 2001; Kantor and Rhodes 1994). Although the fecal streptococci comprise less than 0.1 percent of the digestive tract microflora, this group has been considered to be an excellent indicator of human and animal fecal pollution (Holdeman et al. 1976; WHO 2001). The presence of fecal streptococci correlates well with fecal coliform abundance, and fecal streptococci do not grow in seawater (Geldreich and Kenner 1969; Rozen and Belkin 2001).

The use of the genus *Enterococcus* has received significant attention as a water quality indicator of human bacterial pollution and it is routinely replacing, or at least used in conjunction with, standard coliform monitoring programs. The U.S. EPA (1986) has recommended a geometric mean standard of 35 per 100 ml (5 samples over a 30-day period) and an instantaneous standard of 104 per 100 ml as regulatory standards. However, these recommendations have not been accepted by all of the states, and therefore current *Enterococcus* standards vary. Both the U.S. EPA and the International Shellfish Safety Committee are currently considering the adoption of *Enterococcus* as a regulatory indicator for shellfish-harvesting waters. Several studies have demonstrated that enterococci are better indicators of swimming-associated gastrointestinal illness than are coliform bacteria (e.g., Cabelli et al. 1983; Pruss 1998).

The ratio of fecal coliform to fecal streptococci has been proposed as a means of discriminating between animal and human fecal contamination (Geldreich and Kenner 1969; Sinton et al. 1993; Jagals et al. 1995). Ratios of 4 or greater are generally indicative of human pollution sources, whereas values less than 1 are consistent with animal pollution sources. However, several studies have found that these ratios are not always reliable predictors of human versus animal sources (McFeters et al. 1974; Wheater et al. 1979; Wolfe 1995). Wolfe (1995), in a study of urbanized Murrells Inlet and pristine North Inlet in South Carolina (U.S.A.), found that ratios of fecal coliform to fecal streptococci were <1 at sites with both known human (e.g., septic tanks) and animal pollution sources. The differential mortality rate of fecal coliform and fecal streptococci in seawater, the variable nature of pollution sources in areas affected by nonpoint source runoff, and the lack of consistent results obtained with different growth media make the use of fecal streptococci as an indicator tenuous at best, unless the fecal contamination is very recent (Howell et al. 1995).

Streptococcus bovis is a primary fecal streptococcus bacterium that is found in warm-blooded animals and in birds (Kenner et al. 1960; Clausen et al. 1977; Kantor and Rhodes 1994). *S. bovis* survives poorly in fresh and saline waters compared to fecal coliform bacteria, suggesting that its detection may be indicative of bacterial contamination of recent origin (Kantor and Rhodes 1994). The poor survival in receiving waters and its relatively low prevalence in human feces makes this organism an excellent candidate indicator of contemporary animal bacterial pollution, thus several studies have evaluated the use of *S. bovis* as an indicator system, including studies in southeastern U.S. watersheds. For example, in a Virginia watershed where nonhuman contamination sources were considered the primary sources of fecal materials, *S. bovis* was found throughout the watershed, while other human fecal indicators were not detected (Kantor and Rhodes 1996). Studies such as this suggest the potential utility of this organism as an animal fecal contamination indicator.

11.2.2 Bifidobacteria

Bifidobacteria are obligate, nonmotile, Gram-positive bacteria that have a unique pleomorphic, branching cell morphology (Kantor and Rhodes 1994). The genus *Bifidobacterium* is the third most numerous bacterial population in the human intestine (Charteris et al. 1997) and is the predominant population during childhood (Gibson and Roberfroid 1995). Because of its association with human fecal material, its inability to grow in both fresh and saline waters, and its relative ease of cultivation (Nerbra and Blanch 1999), it has been proposed that *Bifidobacterium* could be used as an indicator of human fecal contamination (Evison and James 1974; Lim et al. 1995), although this proposal has been considered to be controversial (Scardovi et al. 1971; Levin 1977; Cabelli 1978; Resnick et al. 1981). A unique feature of bifidobacteria is their apparent absence and/or low prevalence in animal feces, which may make them useful for differentiating human from animal contamination sources (Levin 1977; Kantor and Rhodes 1994). Physiological differences in sorbitol-fermenting capabilities of bifidobacteria may be used to differentiate human (positive sorbitol fermentor) from animal pollution sources (Mara and Oragui 1983). To date, most studies of bifidobacteria have been conducted in freshwater environments (Carillo et al. 1985; Muñoa and Pares 1988) although the use of bifidobacteria as indicators in marine and estuarine shellfish-harvesting waters has been successfully applied (Cabelli 1978; Kantor and Rhodes 1996).

11.2.3 *Clostridium perfringens*

Prior to the investigation of bifidobacteria as fecal indicators, *Clostridium perfringens* was the only obligatory anaerobic, enteric microorganism considered as a possible indicator of the sanitary quality of water. The anaerobic sulphite-reducing clostridia (SRC), including the organism *C. perfringens,* are much less prevalent than bifidobacteria in human feces, but their spore-forming ability allows them to survive for significant time periods outside the enteric environment in aquatic and estuarine receiving waters (Cabelli 1978). In fact, spores can remain viable in sediments for thousands of years (Edberg et al. 1997). *C. perfringens* is the species of clostridia most often associated with the feces of warm-blooded animals, but is only present in 13–55 percent of human feces (Rosebury 1962; Fujioka et al. 1997). *C. perfringens* is an accepted bacterial indicator of fecal contamination of water (Fujioka and Shizumura 1985; APHA 1992) and in several studies it has been shown to be useful as an indicator of the protozoan enteric pathogen *Giardia* (Ferguson et al. 1996). The usefulness of *C. perfringens* in aquatic systems as indicators of fecal contamination has been explored in several aquatic and estuarine systems and found to be quite useful (e.g., Bezirtzoglou et al. 1994). In addition to having utility as an indicator of contemporary contamination, because of its persistence in sediments, it may be possible to utilize *C. perfringens* to reconstruct urban pollution histories from sediment records.

11.2.4 Bacteriophages

Because coliform bacteria are short lived in estuarine waters they are not generally thought to be predictive of viral contamination in these environments (Feachem et al. 1983; Richards 1985; Kantor and Rhodes 1994). While direct viral measurements in estuarine surface waters are possible (Richards 1985), cost, methodological, and time constraints have prevented their routine use in assessing fecal pollution. Viruses that infect bacteria, known as bacteriophages, can be easily propagated on bacterial hosts, and thus may provide useful surrogate viral indicators of viral and protozoan fecal pollution in estuarine waters. The use of phages as models for indicating the likely presence of pathogens first appeared in the literature in the 1930s, and direct correlation between the presence of certain bacteriophages and the intensity of fecal contamination has been observed in many studies since then. The ease of detection, short assay period, low analytical cost, and resistance to disinfection make bacteriophages potentially useful fecal pollution indicators (Keswisk et al. 1985; DeBartolomeis 1988; Kantor and Rhodes 1994).

The male-specific RNA coliphage (FRNA phage) has been proposed as a potential useful sewage pollution indicator, as warm-blooded animals and sewage are the major sources of these phages (Furuse et al. 1978; Kantor and Rhodes 1994). FRNA phages share many traits with human enteroviruses and may be pollution source specific (human versus animal). FRNA phages comprise from 10–90 percent of total coliphage abundance in domestic waste and sewage (Furuse et al. 1978; Osawa et al. 1981; Kantor and Rhodes 1994). There are four major serological groups (I-IV) that have been identified and have been associated with either animal (Group I), human (Groups II and III), or both animal and human sources (Group IV) (Osawa et al. 1981). Kantor and Rhodes (1996) considered Group IV indicative of animal pollution sources in samples from a tidal creek in Virginia, U.S.A. Similarly, Kantor and Rhodes (1991) measured FRNA phage activity in surface waters and sediments along a salinity gradient in estuarine waters with known human bacterial pollution sources. They found that FRNA phages were infrequently detected in surface waters but were detected at high levels in sediments. Comparisons of FRNA phage and fecal coliform levels in surface water and sediments found very little correlation between their distributions. Other studies have utilized naturally occurring marine bacteriophages as tracers of contamination sources (Paul et al. 1995).

11.2.5 Other Organisms

A number of other groups of naturally occurring aquatic microbes, including *Pseudomonas aeruginosa, Aeromonas hydrophilia, Vibrio,* and *Klebsiella* species have been investigated as indicators of fecal pollution in coastal environments. (e.g., see Borrego et al. 1990; Okpokwasili and Akujobi 1996; Edberg et al. 1997). All of these organisms have characteristics that make them suitable as indicator species of fecal contamination of estuarine and coastal systems. Specifically, studies have indicated that they are well represented in water that has been contaminated, and they are relatively amendable to laboratory test procedures. However, because the growth of these organisms in natural systems is influenced by a large number of unrelated environmental factors, unambiguous interpretation of these indicator strains is difficult. Thus, although significant correlation between these organisms and other fecal indicators has been demonstrated in some cases, their value as general indicators of fecal contamination in coastal systems is largely unknown.

11.3 Microbial Indicators of Eutrophication

Because of their central role in the utilization and remineralization of nutrients, bacteria respond rapidly to increased nutrient loading, particularly organic nitrogen (N), phosphorus (P), and labile carbon. Consequently, monitoring of total bacterial abundance (biomass) appears to be a good indicator of eutrophication of coastal and estuarine waters. A large number of studies have demonstrated a strong relationship between eutrophication, bacterial activity, and total heterotrophic bacteria abundance. This relationship is well illustrated by a long-term monitoring program (1960s to the present) in the western Baltic and most recently discussed in a review by Gocke and Rheinheimer (1991). In this series of studies, among other parameters, total heterotrophic bacteria and bacterial production were monitored in two adjacent fjords draining into the Kiel Bight. One fjord (Kiel Fjord) is surrounded by an urban center, while the other (Schlei Fjord) is bounded by intensely used agricultural areas. While discharges of organic material (sewage) into Kiel Fjord are tightly controlled, Schlei Fjord receives considerable nutrient (inorganic and organic) inputs from nonpoint sources associated with agriculture and is considered to be highly eutrophic. Although similar in morphology, these two estuarine systems have remarkably different characteristics with respect to total heterotrophic bacterial counts and bacterial productivity. Superimposed on variations associated with seasonal cycles, Schlei Fjord consistently supports significantly larger bacterial populations than does Kiel Fjord. For example, summer annual heterotrophic bacterial counts in Kiel Fjord are $3–5 \times 10^6$ cells per ml, while summer bacterial counts in Schlei Fjord are $6–14 \times 10^6$ cells per ml. Likewise, although total bacterial numbers decrease in both fjords during the winter months, Schlei Fjord consistently supports higher bacterial numbers ($2.6–3.5 \times 10^6$ cells per ml) than does Kiel Fjord (0.96×10^6 cells per ml) during the winter. Similarly, large differences in bacterial productivity are observed between these two fjords.

Comparable studies have been conducted in other estuarine systems, including in the southeastern United States, although on less comprehensive scales. For example, high bacterial counts are routinely measured in coastal Georgia, and bacterial counts of up to 3.2×10^7 cells per ml have been reported (Pedrós-Alió and Newell 1989). In the Skidaway River estuary (Savannah, Georgia, U.S.A.) during the period from 1986 to 1996, total bacterial abundance increased nearly 37 percent, from $\sim 5 \times 10^6$ to 8×10^6 cells per ml during the summertime (Figure 11.1).

This increase is correlated with increased development and population on Skidaway Island and with increases in nutrient concentrations, particularly dissolved organic nitrogen (Verity 2002a, 2002b). Effects of eutrophication from the swine and poultry industries also result in increases in total heterotrophic bacteria (including coliform bacteria) and bacterial productivity indices (Burkholder et al. 1997; Mallin et al. 1997), whereas there is some evidence that other types of pollution (e.g., heavy metal contamination) may decrease total bacterial numbers (Schwinghamer 1988). These studies suggest that long-term monitoring of total bacterial biomass can be used as a reliable means of estimating eutrophication impacts and recovery following mitigation

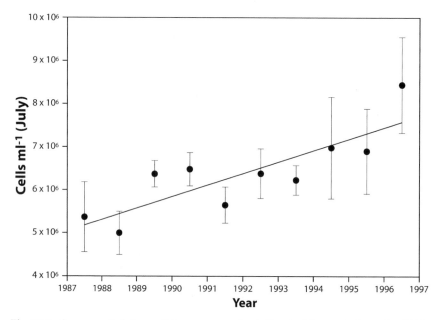

Fig. 11.1. Summer (July) monthly averages of total bacterial counts from the Skidaway River estuary from July 1987 through July 1996. Bacterial abundance increased from 5.3 to 8.4 $\times$ 10^6 cells/ml ($r^2 = 0.70$) over this period. Samples were collected weekly during high tides and counted by epifluoresence microscopy after staining with DAPI. Error bars represent standard deviations of at least three samples during a one-month period. Comparative hydrographic, nutrient, and biological data are available in Verity (2002a, b).

efforts. The simplicity of these methods allows them to be used routinely. However, effects of normal variability associated with seasonal patterns and other site-specific mitigating influences need to be well defined through long-term monitoring if total heterotrophic bacterial densities are to be interpreted with respect to degree of eutrophication or specific ecological impacts.

11.4 Phytoplankton Communities as Indicators

Eutrophication of coastal watersheds has been underway for decades as coastal populations have expanded and land use patterns increasingly reflected anthropogenic activities. This nutrient loading has been coincident with significant increases in primary productivity and plankton biomass (e.g., Brush 1984; Nixon 1997), a causality confirmed in mesocosm studies (D'Elia et al. 1986; Oviatt et al. 1986). Moreover, there is general consensus that the frequency and magnitude of blooms of both benign and harmful indigenous species of phytoplankton have increased during the most recent decades (Smayda 1990; Hallegraeff 1993). A variety of environmental and ecological impacts are fueled by high primary production resulting from nutrient loading (Hobbie and Cole 1984), the most serious of which is often long-term increases in aerial extent and severity of hypoxia that have been reported from several river-dominated coastal ecosystems (Officer et al. 1984; Justic et al. 1995).

Phytoplankton community composition changes with nutrient fluxes because individual taxa have different nutritional requirements (Tilman 1977; Kilham and Kilham 1984; McCormick and Cairns 1994). Enclosure experiments have confirmed that nutrient enrichment alters the community composition of phytoplankton, favoring smaller eukaryotic and prokaryotic species (Goldman and Stanley 1974; Sanders et al. 1987). Field studies report similar occurrences: for example, in Moriches and Great South Bays (New York), extremely dense populations of chlorophytes developed in waters fertilized by effluent from adjacent duck farms, and these blooms coincided with collapse of the oyster fishery (Ryther 1954). Similar occurrences of abundant populations of phytoflagellates in other meso- and eutrophic waters (e.g., Mahoney and McLaughlin 1977; Verity, unpublished data) imply that elevated nutrient concentrations alter phytoplankton community composition favoring smaller, "less-desirable" species (Smayda 1983). Cyanobacterial blooms, in particular, have been increasingly recognized in estuarine environments as a symptom of cultural eutrophication (Paerl and Zehr 2000). Historically, cyanobacterial blooms have been associated with freshwater systems receiving anthropogenic P inputs (Francis 1878; Fogg 1969; Paerl and Tucker 1995) and it has been demonstrated that there is a strong relationship between total N:P ratios and the prevalence of cyanobacterial bloom genera (Smith 1983). N:P ratios below 20 are associated with the development of cyanobacterial blooms. Interestingly, many human-impacted estuarine and coastal waters exhibit N:P ratios well below 20 (D'Elia et al. 1986; Nixon 1986) and appear to be supporting more frequent cyanobacterial blooms (Niemi 1979; Paerl 1996; Paerl and Millie 1996). Common genera of estuarine cyanobacteria associated with eutrophication include *Anabaena*, *Aphanizomenon*, *Lyngbya*, *Nodularia*, and *Oscillatoria*. More recently, blooms of hazardous algal species have been attributed to severe eutrophication of coastal waters (Anderson et al. 2002).

Because of the relatively predictable response of phytoplankton communities to increased nutrient loading (increased productivity, biomass, and community composition) long-term monitoring of these parameters can be used as a reliable indicator of ongoing eutrophication of coastal waters (Verity et al. 2002 a, b).

11.5 Microbial Indicators of Other Contaminants

Marine luminescent bacteria are a ubiquitous group of diverse heterotrophic organisms (Hastings and Nealson 1981). The suppression of the luminescence phenotype by a large number of contaminants is a well-known phenomenon, and luminescence methods have been used to monitor for various organic and inorganic chemical contaminants in a variety of environments and sample types (e.g., see Steinberg et al. 1995). In pristine marine environments, including estuarine and coastal areas, luminescent marine bacteria can account for as much as 10 percent of the total heterotrophic bacteria present (Ramaiah and Chandramohan 1993). However, in the presence of a wide range of pollutants that result from a variety of different human activities (e.g., heavy metals, organic solvents, detergents, pesticides, alcohols, straight-chained hydrocarbons, PCBs, and PAHs) the number of luminescent bacteria can be greatly depressed. For example, in a recent seven-year study, Ramaiah and Chandramohan (1993) reported the complete absence of luminescent bacteria in more than 200 samples from polluted sites in India, while luminescent bacteria were abundant in an equal number of pristine sites. Studies in the Skidaway River estuary have demonstrated that the frequency of luminescent bacteria is depressed by exposure to hydrocarbons (diesel fuel) and salt marsh sediments contaminated with mercury and PCBs (Figure 11.2), suggesting that monitoring of luminescent bacteria could be used as a primary screening tool to detect impacted coastal environments. However, a number of important caveats must be considered. For example, since there can be considerable normal variability associated with season, salinity, and nutrient sources, the use of luminescence parameters, as with heterotrophic bacteria counting methods, requires that long-term seasonal baseline data sets be developed to utilize the luminescence criteria effectively in monitoring and identifying human-impacted coastal ecosystems. Furthermore, since a large number of types of contamination appear to suppress luminescence of marine bacteria, this method will not necessarily be useful for identifying the type or source of contaminant. However, as with heterotrophic plate count techniques, determining the fraction of luminescent bacteria is an extremely simple procedure and could therefore be routinely accomplished in long-term monitoring studies. Simplicity and sensitivity of detection and the ubiquity of luminescent bacteria in coastal waters make this method an attractive candidate as an alternative microbial indicator of marine ecosystem health.

In addition to taking advantage of naturally occurring bioluminescent bacterial communities, a number of assays utilize specific cultured bioluminescent species as an assay to assess the toxicity of chemical contaminants in coastal environments. For example, the Microtox™ assay utilizes the marine bacterium, *Vibrio fisheri*. In the presence of chemical contaminants, the respiration rate of *V. fisheri* is retarded and luminescence is inhibited. Long et al. (1998) used this technique to assess the toxicity of sediment-bound contaminants from Winyah Bay, Charleston Harbor, and Leadenway Creek in South Carolina, and the Savannah River and Brunswick Harbor estuaries in Georgia. Results from these studies suggested that each of these environments was impacted relative to appropriate pristine reference sites. A variant on this approach is the Mutatox™ assay, in which a dark variant of *V. fisheri* is used to assay mutagenesis

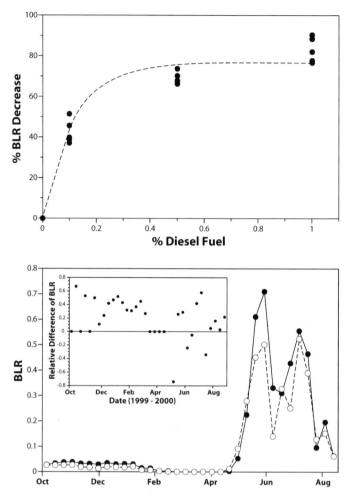

Fig. 11.2. Effect of contaminant exposure on the Bioluminescent Ratio (BLR) of indigenous bacteria from the Skidaway River estuary. (A) The impact of diesel fuel. BLR decreased exponentially in response to increased amount of diesel fuel in experimental incubations. (B) Impact of exposure to pristine and contaminated salt marsh sediments. Exposure to mercury (Hg) and polychlorinated biphenyl (PCB) contaminated salt marsh sediments typically resulted in a decrease in the BLR of indigenous bacterial populations from the Skidaway River estuary. (Inset) The BLR was higher in water samples collected after passage through uncontaminated vs. contaminated sediments in 22 of 27 (82 percent) weekly samples collected from October 1999 through August 2000. Comparisons were facilitated using a salt marsh mesocosm facility. Modified from Frischer et al. (2005), with permission from ASA-CSSA-SSSA.

(Johnson and Long 1998). Mutagenic chemicals cause this dark variant to mutate to a bioluminescent form and increased light output is related to mutagenicity.

11.6 Other Indicator Bacteria

Indigenous microbial communities are composed of diverse assemblages of microorganisms. Inputs of specific contaminants associated with human activities may result in changes in the taxonomic composition of coastal microbial communities (bacteria and phytoplankton). If such changes were consistently associated with particular contaminants associated with human activity, then these organisms or measures of community composition might be used as indicators of these inputs. As new techniques are developed to rapidly identify specific bacteria or to fingerprint the composition and structure of microbial communities (e.g., molecular probes, genetic fingerprinting, DNA microarrays, etc.), it is becoming increasingly possible to monitor a larger number of specific organisms in complex microbial consortia. A full description of these emerging techniques and approaches is beyond the scope of this chapter. However, researchers have begun to investigate the effect of various types of contamination on the abundance of specific microorganisms and communities as indicators of coastal ecosystem health. Using such techniques, a number of bacterial species can now be relatively simply monitored and are currently undergoing evaluation as indicators of human impact on coastal systems. Such bacteria include a number of *Pseudomonas* and *Acinetobacter* species and other taxa including *Burkholderia cepacia, Bacillus, Myxococcus,* and *Chromobacterium* species (see Table 11.2). Unfortunately, the results of these studies are not always unambiguous, and thus it remains unknown whether these species can be applied generally as indicator species or if they will be limited to specific sites under carefully defined and constrained conditions. Nonetheless, as the practical ability to rapidly identify bacteria at finer taxonomic levels (eventually to specific genes and molecules) improves, it may be that useful indicator species of ecosystem function will be discovered. Alternatively, since ecosystem function is ultimately an emergent property of complex interactions between diverse populations and environmental variables, it is also possible that a single universal indicator species does not exist and therefore monitoring the effects of human activity on ecosystem health will require more complex indicator criteria than individual species.

11.7 Microbial Community Composition as an Indicator of Coastal Ecosystem Health

The functioning of microbial systems depends on complex interactions among bacteria and phytoplankton; abiotic and biotic conditions of the environment; mechanisms of metabolic regulation and adaptation; and mutual relationships between the microorganisms and macroorganisms. Because of the theoretically large number of variables that can influence microbial communities, these variables likely lead to the establishment of unique communities that reflect all of these interactions. Therefore, changes in environmental conditions as a result of human activities are expected to influence not only the functioning of microbial communities, but also their composition and structure. A major question that remains unanswered is the extent to which these changes in composition are reproducible and predictable, and can therefore be used to assess and monitor land use effects.

The development of molecular tools and approaches for identification and localization of microorganisms has resulted in a profound shift in our understanding of microbial communities. Even when microbiologists were exclusively dependent upon culturing techniques and microscopy to investigate the distribution of microorganisms, it was recognized that most microorganisms in natural habitats were not being analyzed. This was because the microorganisms that could be observed and counted directly by microscopy exceeded by 10- to 100-fold those that could be cultured in the laboratory. Thus, much of our understanding of the physiology, taxonomy, and ecology of aquatic organisms was based on a small proportion of the total populations. The advent of molecular approaches in the last two decades has opened the door to the detection and identification of microorganisms and microbial communities that could not be cultured, and more recently, methods are being developed to study their physiology and ecology in situ. Despite this progress, we are only just beginning to develop methods for characterizing total community structure and to relate that structure to marine environmental determinants; recent reviews summarize molecular approaches that are under development to investigate microbial communities (Theron and Cloete 2000; Stevens et al. 2001; Zhou 2003).

11.8 Conclusion

Coliform bacteria (total and fecal) have been used for nearly a century as primary bacterial indicators of water quality. However, these organisms are not necessarily appropriate for use in coastal systems or as a general gauge of "ecosystem health." Recently, enterococci bacteria have gained general acceptance over coliform bacteria as appropriate indicators of sewage contamination in estuarine systems, particularly in coastal bathing waters. A number of other bacterial species or groups of bacteria, other than coliform or enterococci bacteria, have also been examined as water quality indicators in coastal areas. As methodological advances facilitating the simple and unambiguous detection of these species in coastal waters occur, it is becoming increasingly feasible to utilize these alternative organisms as water quality criteria in coastal and estuarine environments. Additionally, new direct methods, independent of cultivation procedures, are also leading to the possibility of routine direct detection of pathogens and may eventually obviate the need for indicators altogether. However, although these organisms and techniques show great promise, none have been widely adopted by the regulatory or management communities.

In addition to reliable indicators of human pathogens, there is also a considerable interest in the identification of indicators of ecosystem health. Microorganisms, which serve as the primary regenerators of nutrients and play important role in marine food webs, are logical choices as ecosystem indicators. Eutrophication of coastal systems has been recognized as a primary consequence of human activities, and a large number of studies have convincingly demonstrated that both bacterial and phytoplankton biomass and activity respond to increased nutrient loading. Thus, long-term time series measurements of bacterial and phytoplankton can be used as reliable indicators of eutrophication in both marine and freshwater systems. Furthermore, the type of organisms present in waters undergoing eutrophication can be used as an indicator of the eutrophication process. For example, the dominance of smaller flagellated phytoplankton and cyanobacteria species is typically associated with eutrophic waters. The presence of bioluminescent bacteria, specifically the ratio of bioluminescent to nonbioluminescent bacteria in marine systems, also appears to be a good candidate as a general indicator of estuarine water quality. However, effects of normal variability associated with seasonal patterns and other site-specific mitigating influences need to be well defined before these approaches can be routinely utilized.

Because the functioning of all natural ecosystems depends on healthy microbial communities, the structures of these communities themselves are likely to be the ultimate indicators of ecosystem health and water quality in the future. Considerable research and technique development is underway to develop approaches for defining microbial community structure and relating these measures to ecosystem function. Progress, specifically the development of gene-based approaches, will likely facilitate the direct measurement of microbial community structure in the relatively near future.

Acknowledgments

This work was supported by the U.S. National Oceanic and Atmospheric Administration through the Land Use-Coastal Ecosystem Study (LUCES) State of Knowledge program (NOAA Coastal Ocean Program grant NA960PO113). Other support was from the U.S. National Science Foundation (OPP–00–83381 and OCE 99–82133), and the U.S. Department of Energy (FG02–88ER62531 and FG02–98ER62531). We also wish to acknowledge the work of several students and technical staff who have participated in our investigations of the Skidaway River estuary over the years, including: Victoria Baylor, Anke Blechschmidt, Jean Danforth, Susan Ebanks, Tara Foy, Daphne Heaton, Ronnie Juraske, Sandra Pagen, Tina Walters, Samantha Williams, and Ying Hong.

References

Anderson, D. M., P. M. Glibert, and J. M. Burkholder. 2002. Harmful algal blooms and eutrophication: nutrient sources, composition, and consequences. *Estuaries* 25:704–726.

American Public Health Association. 1992. Standard Methods for the Examination of Water and Wastewater, 18th ed. American Public Health Association, Washington, DC. 947 p.

Ashbolt, N. J., C. S. Kueh, and G. S. Grohmann. 1993. Significance of specific bacterial pathogens in the assessment of polluted receiving water of Sydney, Australia. *Water Science and Technology* 27:449–452.

Bezirtzoglou, E., D. Dimitriou, A. Panagiou, I. Kagalou, and Y. Demoliates. 1994. Distribution of *Clostridium perfringens* in different aquatic environments in Greece. *Microbiological Research* 149:129–134.

Borrego, J. J., R. Cornax, M. A. Morinigo, E. Martinez-Manzanares, and P. Romero. 1990. Coliphages as an indicator of faecal pollution in water. Their survival and productive infectivity in natural aquatic environments. *Water Research* 24:111–116.

Brush, G. S. 1984. Stratigraphic evidence of eutrophication in an estuary. *Water Research* 20:531–541.

Burkholder, J. M., M. A. Mallin, H. B. Glasgow Jr., L. M. Larsen, M. R. McIver, G. C. Shank, N. Deamer-Melia, D. S. Briley, J. Springer, B. W. Touchette, and E. K. Hannon. 1997. Impacts to a coastal river and estuary from rupture of a large swine waste holding lagoon. *Journal of Environmental Quality* 26:1451–1466.

Cabelli, V. J., A. P. Dufour, L. J. McCabe, and M.A. Levin. 1983. A marine recreational water quality criterion consistent with indicator concepts and risk analysis. *Journal-Water Pollution Control Federation* 55:1306–1314.

Cabelli, V. J. 1978. New standards for enteric bacteria, pp. 233–271. In, Mitchell, R. (ed.), Water Pollution Microbiology, Vol. 2. Wiley, New York, NY. 422 p.

Cabelli, V. J. 1983. Health effects criteria for marine recreational waters. EPA-600/1–80–031, U.S. Environmental Protection Agency, Research Triangle Park, NC. 105 p.

Carillo, M., E. Estrada, and T. C. Hazen. 1985. Survival and enumeration of fecal coliform indicators *Bifidobacterium adolescentis* and *Esherichia coli* in a tropical rain forest watershed. *Applied Environmental Microbiology* 50: 468–476.

Carson, C. A., B. L. Shear, M. R. Ellersieck, and A. Asfaw. 2001. Identification of fecal *Escherichia coli* from humans and animals by ribotyping. *Applied Environmental Microbiology* 67:1503–1507.

Charteris, W. P., P. M. Kelly, L. Morelli, and J. K. Collins. 1997. Selective detection, enumeration and identification of potentially probiotic *Lactobacillus* and *Bifidobacterium* species in mixed bacterial populations. *International Journal of Food Microbiology* 35:1–27.

Clausen, E. M., B. L. Green, and W. Litsky. 1977. Fecal streptococci: Indicators of pollution, pp. 247–264. In, Hoadley, A. W. and B. J. Dutka (eds.), Bacterial Indicators/Health Hazards Associated with Water. American Society for Testing Materials, Special Technical Publication #635, Philadelphia, PA.

DeBartolomeis, J. 1988. The enumeration of F-specific bacteriophages from environmental waters. Ph.D. thesis. University. of Rhode Island, Kingston, RI. 111 p.

D'Elia, C. F., J. G. Sanders, and W. R. Boyton. 1986. Nutrient enrichment studies of a coastal plain estuary: phytoplankton growth in large-scale continuous cultures. *Canadian Journal of Fisheries and Aquatic Science* 43:397–406.

Edberg, S. C., H. LeClerc, and J. Robertson. 1997. Natural protection of spring and well drinking water against surface microbial contamination. II. Indicators and monitoring parameters for parasites. *Critical Reviews in Microbiology* 23:179–206.

Edberg, S. C., E. W. Rice, R. J. Karlin, and M. J. Allen. 2000. *Escherichia coli:* The best biological drinking water indicator for public health protection. *Journal of Applied Microbiology—Symposium Supplement* 88:106S–116S.

Evison, L. M. and A. James. 1974. *Bifidobacterium* as an indicator of faecal pollution in water, pp. 107–116. In, Proceedings of the 7th International Conference on Water Pollution Research, Paris, September 9–13, 1974. Pergamon Press, Oxford, UK.

Feachem, R. G., D. J. Bradley, H. Garelick, and D. D. Mara. 1983. Sanitation and disease. Health aspects of excreta and wastewater management. World Bank Studies in Water Supply and Sanitation. Wiley, New York, NY. 534 p.

Ferguson, C. M., B. G. Coote, N. J. Ashbolt, and I. M. Stevenson. 1996. Relationships between indicators, pathogens and water quality in an estuarine system. *Water Research* 30:2045–2054.

Fletcher, M., P. G. Verity, M. E. Frischer, K. A. Maruya, and G. I. Scott. 1998. Microbial indicators and phytoplankton and bacterial indicators as evidence of contamination caused by changing land use patterns. Land Use-Coastal Ecosystem Study State of Knowledge Report, South Carolina Sea Grant Consortium, Charleston, SC. 83 p. http:\ \ www.lu-ces.org \ documents \ sokreports \ microbialindicators.pdf

Fogg, G. E. 1969. The physiology of an algal nuisance. *Proceedings of the Royal Society of London Series B* 173:175–189.

Francis, G. 1878. Poisonous Australian lake. *Nature* 18:11–12.

Frischer, M .E., J. M. Danforth, M. A. Newton Healy, and F. M. Saunders. 2000. Whole cell versus total RNA extraction for the analysis of microbial community structure using 16S rRNA targeted oligonucleotide probes in saltmarsh sediments. *Applied Environmental Microbiology* 66:3037–3043.

Frischer, M. E., J. M. Danforth, T. F. Foy, and R. Juraske. (2005.) Bioluminescent bacteria as indicators of human impact in coastal estuarine systems. *Journal of Environmental Quality.* 34:1328–1336.

Fujioka, R. S. and L. K. Shizumura. 1985. *Clostridium perfringens,* a
reliable indicator of stream water quality. *Journal-Water Pollution
Control Federation* 57:986–992.

Fujioka, R. S., B. Roll, and M. Byappanahilli. 1997. Appropriate recreational water quality standards for Hawaii and other tropical
regions based on concentrations of *Clostridium perfringens. Proceedings of the Water Environment Federation, 70th Annual Conference and
Exposition* 4:406–411.

Furuse, K., T. Sukaraa, A. Hirashima, M. Katsuki, A. Ando and
I. Watanabe. 1978. Distribution of ribonucleic acid coliphages in south
and east Asia. *Applied Environmental Microbiology* 35:995–1002.

Geldreich, E. E. and B. A. Kenner. 1969. Concepts of fecal streptococci
in stream pollution. *Journal-Water Pollution Control Federation*
41:336–352.

Gibson, G. R. and M. B. Roberfroid. 1995. Dietary modulation of the
human colonic microbiota: introducing the concept of prebiotics.
Journal of Nutrition 125:1401–1402.

Gocke K. and G. Rheinheimer. 1991. Influence of eutrophication on
bacteria in two fjords of the western Baltic. *International Review of
Hydrobiology* 76:371–385.

Goldman, J. C. and H. I Stanley. 1974. Relative growth of different species of marine algal in wastewater-seawater mixtures. *Marine Biology* 28:17–25.

Grabow, W. O. K., T. E. Neubrech, C. S. Holzhausen, and J. Jofre.
1995. *Bacteroides fragilis* and *Escherichia coli* bacteriophages:
Excretion by humans and animals. *Water Science and Technology*
31:223–230.

Guimaraes, V. F., M. A. V. Araujo, C. S. Leda, M. Hagler, and A. N. Hagler.
1993. *Pseudomonas aeruginosa* and other microbial indicators of
pollution in fresh and marine waters of Rio de Janeiro, Brazil.
Environmental Toxicology and Water Quality 8:313–322.

Hallegraeff, G. M. 1993. A review of harmful algal blooms and their
apparent global increase. *Phycologia* 32:79–99.

Hastings, J. W. and K. H. Nealson. 1981. The symbiotic luminous bacteria, pp. 1332–1345. In, Starr, M. P., H. Stolp, H. G. Truper, A. Balows,
and H. G. Schlegel (eds.), The Procaryotes. Springer-Verlag,
New York, NY.

Hobbie, J. E. and J. J. Cole. 1984. Response of a detrital foodweb to eutrophication. *Bulletin of Marine Science* 35:357–363.

Holdeman, L. V., I. J. Good, and W. E. C. Moore. 1976. Human fecal flora: Variation in bacterial composition within individuals and a possible effect of emotional stress. *Applied Environmental Microbiology* 31:359–375.

Howell, J. M., M. S. Coyne, and P. L. Cornelius. 1995. Faecal bacteria in agricultural waters of the blue grass region of Kentucky. *Journal of Environmental Quality* 24:411–419.

Kantor, H. and M. Rhodes. 1991. Indicators and alternative indicators of growing water quality, pp. 135–196. In, Ward, D. R. and C. R. Hackney (eds.), Microbiology of Marine Food Products. Van Nostrand Rheinhold, New York, NY.

Jagals, P. and W. O. K. Grabow. 1996. An evaluation of sorbitol-fermenting bifidobacteria as specific indicators of human faecal pollution of environmental water. *Water SA* 22:235–238.

Jagals, P. 1997. Stormwater runoff from typical developed and developing South African urban developments: definitely not for swimming. *Water Science Technology* 35:133–140.

Jagals, P. W., O. K. Grabow, and J. C. deVilliers. 1995. Evaluation of indicators for assessment of human and animal faecal pollution of surface run-off. *Water Science Technology* 31: 235–241.

Jamieson, R. C., R. J. Gordon, S. C. Tattrie, and G. W. Stratton. 2003. Sources and persistence of fecal coliform bacteria in a rural watershed. *Water Quality Research Journal of Canada* 38:33–47.

Johnson, B. T. and E. R. Long. 1998. Rapid toxicity assessment of sediments from estuarine ecosystems: a new tandem *in vitro* testing approach. *Environmental Toxicology and Chemistry* 17:1099–1106.

Justic, D., N. N. Rabalais, and R. E. Turner. 1995. Stoichiometric nutrient balance and origin of coastal eutrophication. *Marine Pollution Bulletin* 30:41–46.

Kantor, H. and M. Rhodes. 1994. Microbial and chemical indicators, pp. 30–91. In, Hackney, C. M. and M. D. Pierson (eds.), Environmental Indicators and Shellfish Safety. Chapman and Hall, New York, NY.

Kantor, H. and M. Rhodes. 1996. Identification of pollutant sources contributing to degraded sanitary water quality in Taskinas Creek National Estuarine Research Reserve, Virginia. Final Report, Taskinas National Estuarine Research Reserve, OCRM #NA47OR0199, College of William and Mary, VIMS, Gloucester Point, VA. 47 p. + Appendices.

Kaspar, C. W. and M. L. Tamplin. 1993. Effects of temperature and salinity on the survival of *Vibrio vulnificus* in seawater and shellfish. *Applied Environmental Microbiology* 59:2425–2429.

Kenner, B. A., H. F. Clark, and P. Kabler. 1960. Fecal streptococci. II. Quantification of streptococci in feces. *American Journal of Public Health* 50:1553–1559.

Keswisk, B. H., T. S. Satterwhite, P. C. Johnson, H. L. DuPont, S. L. Secor, J. A. Bitsura, G. W. Gary, and J. C. Hoff. 1985. Inactivation of Norwalk virus in drinking water by chlorine. *Applied Environmental Microbiology* 50:261–264.

Kilham, S. S. and P. Kilham. 1984. The importance of resource supply rates in determining phytoplankton community structure, pp. 7–27. In, Meyers, D. G. and J. R. Strickler (eds.), Trophic Interactions Within Aquatic Ecosystems. Westview Press, Boulder, CO.

Lemke, M. J., B. J. Brown, and L. G. Leff. 1997. The response of three bacterial populations to pollution in a stream. *Microbial Ecology* 34:224–231.

Levin, M. A. 1977. Bifidobacteria as water quality indicators, pp. 131–138. In, Hoadley, A. W. and B. J. Dutka (eds.), Bacterial Indicators/Health Hazards Associated with Water. American Society for Testing Materials, Special Technical Publication #635, Philadelphia, PA.

Lim, K. S., C. S. Huh, and Y. J. Baek. 1995. A selective enumeration medium for bifidobacteria in fermented dairy products. *Journal of Dairy Science* 78:2108–2112.

Long, E. R., G. I. Scott, J. Kucklick, M. H. Fulton, B. C. Thompson, R. S. Carr, J. Beidenbach, K. J. Scott, G. B. Thurby, G. T. Chandler, J. W. Anderson, and G. M. Sloane. 1998. Magnitude and extent of sediment toxicity in selected estuaries of South Carolina and Georgia. Technical Memorandum NOS ORCA 128. NOAA/NOS, U.S. Department of Commerce, Washington, DC. 289 p.

Lucena, F., R. Araujo, and J. Jofre. 1996. Usefulness of bacteriophages infecting *Bacteroides fragilis* as index microorganisms of remote faecal pollution. *Water Research* 30:2812–2816.

Mahoney, J. B. and J. J. A. McLaughlin. 1977. The association of phytoflagellates in lower New York Bay with hypereutrophication. *Journal of Experimental Marine Biology and Ecology* 28:53–65.

Mallin, M. A., J. M. Burkholder, M. R. McIver, G. C. Shank, H. B. Glasgow Jr., B. W. Touchette, and J. Springer. 1997. Comparative effects of poultry and swine waste lagoon spills on the quality of receiving streamwaters. *Journal of Environmental Quality* 26:1622–1631.

Mara, D. D. and J. I. Oragui. 1983. Bacteriological methods for distinguishing between human and animal faecal pollution of water: results of fieldwork in Nigeria and Zimbabwe. *Bulletin of the World Health Organization* 63:773–783.

Mara, D. D. and J. I. Oragui. 1985. Sorbital fermenting bifidobacteria as specific indicators of human fecal pollution. *Journal of Applied Bacteriology* 55:349–357.

Madsen, E. L., A. Winding, K. Malachowsky, C. T. Thomas, and W. C. Ghiorse. 1992. Contrasts between subsurface microbial communities and their metabolic adaptation to polycyclic aromatic hydrocarbons at a forested and an urban coal-tar disposal site. *Microbial Ecology* 24:199–213.

McCormick, P. V. and J. Cairns, Jr. 1994. Algae as indicators of environmental change. *Journal of Applied Phycology* 6:509–526.

McFeters, G. A., G. K. Bissonette, J. J. Jezeski, C. A. Thomson, and D. G. Stuart. 1974. Comparative survival of indicator bacteria and enteric pathogens in well water. *Applied Environmental Microbiology* 27:823–829.

Mezrioui, N., K. Oufdou, and B. Baleux. 1995. Dynamics of non-O1 *Vibrio cholerae* and fecal coliforms in experimental stabilization ponds in the arid region of Marrakesh, Morocco, and the effect of pH, temperature, and sunlight on their experimental survival. *Canadian Journal of Microbiology* 41:489–498.

Milner, C. R. and R. Goulder. 1985. The abundance, heterotrophic activity and taxonomy of bacteria in a stream subject to pollution by chlorophenols, nitrophenols and phenoxyalkanoic acids. *Water Research* 20:85–90.

Muñoa, F. J. and P. Pares. 1988. Selective medium for isolation and enumeration of *Bifidobacterium* spp. *Applied Environmental Microbiology* 54:1715–1718.

Nebra, Y. and A. R. Blanch. 1999. A new selective media for *Bifidobacterium* spp. *Applied Environmental Microbiology* 65:5173–5176.

Niemi, A. 1979. Blue-green algal blooms and N:P ratio in the Baltic Sea. *Acta Botonica Fennica* 110:57–61.

Nixon, S. W. 1986. Nutrient dynamics and the productivity of marine coastal waters, pp. 97–115. In, Halwagy, R., D. Clayton, and M. Behbehani (eds.), Marine Environment and Pollution. Alden Press, Oxford, UK.

Nixon, S. W. 1997. Prehistoric nutrient inputs and productivity in Narragansett Bay. *Estuaries* 20: 253–261.

Officer, C. B., R. B. Biggs, J. L. Taft, L. E. Cronin, M. Tyler and W. R. Boynton. 1984. Chesapeake Bay anoxia: Origin, development, and significance. *Science* 223:22–27.

Okpokwasili, G. C. and T. C. Akujobi. 1996. Bacteriological indicators of tropical water quality. *Environmental Toxicology and Water Quality* 11:77–81.

Oragui, J. I. and D. D. Mara. 1983. Investigation of the survival characteristics of *Rhodococcus coprophilus* and certain fecal indicator bacteria. *Applied Environmental Microbiology* 46:356–360.

Osawa, S., K. Furuse, and I. Watanabe. 1981. Distribution of ribonucleic acid coliphages in animals. *Applied Environmental Microbiology* 41:164–168.

Oviatt, C. A., A. A. Keller, P. A. Sampou, and L. L. Beatty. 1986. Patterns of productivity during eutrophication: a mesocosm experiment. *Marine Ecology Progress Series* 28:69–80.

Paerl, H. W. 1996. Microscale physiological and ecological studies of aquatic cyanobacteria: Macroscale implications. *Microscopy Research and Technique* 33:47–42.

Paerl, H. W. and D. F. Millie. 1996. Physiological ecology of toxic cyanobacteria. *Phycologia* 35:160–167.

Paerl, H. W. and C. Tucker. 1995. Ecology of blue-green algae in aquaculture ponds. *Journal of the World Aquaculture Society* 26:1–53.

Paerl, H. W., and J. P. Zehr. 2000. Marine nitrogen fixation, pp. 387–426. In, Kirchman, D. L. (ed.), Microbial Ecology of the Oceans. Wiley-Liss, New York, NY.

Paul, J. H., J. B. Rose, J. Brown, E. A. Shinn, S. Miller and S. R. Farrah, S. R. 1995. Viral tracer studies indicate contamination of marine waters by sewage disposal practices in Key Largo, Florida. *Applied Environmental Microbiology* 61:2230–2234.

Pedrós-Alió, C. and S. Y. Newell. 1989. Microautoradiographic study of the thymidine uptake in brackish waters around Sapelo Island, Georgia, USA. *Marine Ecology Progress Series* 55:83–94.

Pinto, B., R. Pierotti, G. Canale, and D. Reali. 1999. Characterization of faecal streptococci as indicators of feacal pollution and distribution in the environment. *Letters in Applied Microbiology* 29:258–263.

Pruss, A. 1998. A review of epidemiological studies from exposure to recreational water. *International Journal of Epidemiology* 27:1–9.

Ramaiah, N. and D. Chandramohan. 1993. Ecological and laboratory studies on the role of luminous bacteria and their luminescence in coastal pollution surveillance. *Marine Pollution Bulletin* 26:190–201.

Resnick, I. G. and M. A. Levin. 1981. Assessment of bifidobacteria as indicators of human fecal pollution. *Applied Environmental Microbiology* 42:433–438.

Rhodes, M. W. and H. Kantor. 1994. Seasonal occurrence of mesophilic *Aeromonas* spp. as a function of biotype and water quality in temperate freshwater lakes. *Water Research* 28:2241–2251.

Richards, G. P. 1985. Outbreaks of shellfish associated enteric virus illness in the United States. Requisite for development of viral guidelines. *Journal of Food Protection* 48:15–23.

Rosebury, T. 1962. Microorganisms Indigenous to Man, pp. 87–90 and 332–335. McGraw-Hill, New York, NY.

Rozen, Y. and S. Belkin. 2001. Survival of enteric bacteria in seawater. *FEMS Microbiology Reviews* 25:513–529.

Ryther, J. H. 1954. The ecology of phytoplankton blooms in Moriches Bay and Great South Bay, Long Island, New York. *Biological Bulletin* 106:198–209.

Sabrilli, G., M. Crusanti, M. Bucci, C. Gaggi, and E. Bacci. 1997. Marine heterotrophic bacteria as indicators in the quality assessment of coastal waters: introducing the "apparent bacterial concentration" approach. *Environmental Toxicology and Chemistry* 16:135–139.

Sanders, J. G., S. J. Cibik, C. F. D'Elia, and W. R. Boynton. 1987. Nutrient enrichment studies in a coastal plain estuary: changes in phytoplankton species composition. *Canadian Journal of Fisheries and Aquatic Sciences* 44:83–90.

Scardovi, V., L. D. Trovatelli, G. Zani, F. Crociani, and D. Matteuzzi. 1971. Deoxyribonucleic acid homology relationships among species of the genus *Bifidobacterium*. *International Journal of Systematic and Evolutionary Microbiology* 21:276–294.

Schwinghamer, P. 1988. Influence of pollution along a natural gradient and in a mesocosm experiment on sediment microbial numbers and biomass. *Marine Ecology Progress Series* 46:193–197.

Sherr E. and B. Sherr. 2000. Marine microbes: An overview, pp. 13–46. In, Kirchman, D. L. (ed.), Microbial Ecology of the Oceans. Wiley-Liss, New York, NY.

Sinha, S. N. and R. D. Banerjee. 1997. Ecological role of thiosulfate and sulfide utilizing purple nonsulfur bacteria of a riverine ecosystem. *FEMS Microbiology Ecology* 24:211–220.

Sinton, L. W., A.M. Donnison, and C. M. Hastie. 1993. Fecal streptococci as fecal pollution indicators—a review 2. Sanitary significance, survival, and use. *New Zealand Journal of Marine and Freshwater Research* 27:117–137.

Smayda, T. J. 1983. The phytoplankton of estuaries, pp. 65–102. In, Ketchum B. H. (ed.), Estuaries and Enclosed Seas. Elsevier, New York, NY.

Smayda, T. J. 1990. Novel and nuisance phytoplankton blooms in the sea: evidence for a global epidemic, pp. 29–40. In, Graneli, E., B. Sundstrom, L. Edler, and D. M. Anderson (eds.), Toxic Marine Phytoplankton. Elsevier, Amsterdam,.

Smith, V. H. 1983. Nitrogen, phosphorus, and nitrogen fixation in lacustrine and estuarine ecosystems. *Limnology and Oceanography* 35:1852–1859.

Snow, J. 1855. On the Mode of Communication of Cholera. John Churchill, London. 38 p.

Steinberg, S. M., E. J. Poziomek, W. H. Engelmann, and K. R. Rogers. 1995. A review of environmental applications of bioluminescence measurements. *Chemosphere* 30:2155–2197.

Stevens, M., N. Ashbolt, and D. Cunliffe. 2003 (cited 2006). Recommendations to change the use of coliforms as microbial indicators of drinking water quality. National Health and Research Council, Canberra, Australia. Available from: http://nhmrc.gov.au/publications/-files/eh32.pdf.

Theron, J. and T. E. Cloete. 2000. Molecular techniques for determining microbial diversity and community structure in natural environments. *Critical Reviews in Microbiology* 26:37–57.

Tilman, D. 1977. Resource competition between planktonic algae: an experimental and theoretical approach. *Ecology* 58:338–348.

Toranzos, G. A. 1991. Current and possible alternate indicators of fecal contamination in tropical waters: A short review. *Environmental Toxicology and Water Quality: An International Journal* 6:121–130.

U.S. Environmental Protection Agency (U.S. EPA). 1986. Ambient water quality criteria for bacteria. EPA440/5–84–002. Office of Water Regulation and Standards, Criteria and Standards Division. U.S. Environmental Protection Agency, Washington, DC. 17 p.

U.S. Public Health Service. 1995. Interstate Shellfish Sanitation Conference, Manual of Operations, Part 1. Sanitation of shellfish-harvesting areas, pp. C1–C29. U.S. Food and Drug Administration, Center for Food Safety and Applied Nutrition, Office of Seafood, Washington, DC.

Verity, P. G. 2002a. A decade of change in the Skidaway River Estuary. I. Hydrography and nutrients. *Estuaries* 25:944–960.

Verity, P. G. 2002b. A decade of change in the Skidaway River Estuary. II. Particulate organic carbon, nitrogen, and chlorophyll a. *Estuaries* 25:961–975.

Wheater, D. W. F., D. D. Mara, and J. Oragui. 1979. Indicator systems to distinguish sewer from stormwater runoff and human from animal fecal material, pp. 1–25. In, James, A. and L. Evison (eds.), Biological Indicators of Water Quality. Wiley, Chichester, UK.

Whitman, R. L. and M. B. Nevers. 2003. Foreshore sand as a source of *Escherichia coli* in nearshore water of a Lake Michigan beach. *Applied Environmental Microbiology* 69:5555–5562.

World Health Organization. 1996. Guidelines for Drinking Water Quality. 2nd ed. Vol. 2. Health criteria and other supporting information. World Health Organization, Geneva. 133 p.

World Health Organization. 2001. Bathing water quality and human health. WHO/SDE/WSH/01.1. Protection of the Human Environment, Water, Sanitation and Health Series. World Health Organization, Geneva.

Wolfe, T. 1995. A comparison of fecal coliform bacterial densities and fluorescent intensities in Murrells Inlet, a highly urbanized estuary and North Inlet, a pristine forested estuary. Master's thesis, School of Public Health, University of South Carolina, Columbia. 84 p.

Zhou, J. H. 2003. Microarrays for bacterial detection and microbial community analysis. *Current Opinion in Microbiology* 6:288–294.

Afterword

Managing Coastal Urbanization and Development in the Twenty-First Century: The Need for a New Paradigm

Geoffrey I. Scott, A. Frederick Holland, and Paul A. Sandifer

The marine environment consists of three major zones: the coastal zone, continental shelf, and open ocean. The effects of most anthropogenic contamination occur in the near-coastal zone which consists of estuaries, bays, sounds, and wetlands. More than 76 percent of all commercially and recreationally important fish and shellfish species are estuarine dependant, usually spending their earliest and most sensitive life history stages in near-coastal zone nursery grounds (Fulton et al. 1996; NOAA 1999). It has only been recognized recently that these areas are not only a resource base of national significance but are also among the nation's most highly stressed natural systems (U.S. EPA 1999). For example, approximately, 44 percent of estuaries assessed in 1998 were impaired (U.S. EPA 2000). The leading sources of pollution in these coastal areas are urban runoff, municipal sewage, agricultural runoff, and industrial wastewater. Additional causes of degradation include shoreline modification, overfishing, and high-density recreational use. Beach closings, fish and shellfish consumption advisories, increasing incidences of red tides, and evidence of toxic substances in sediments and biota are indicative of pollution-related declines in coastal environmental quality.

The Coastal Zone Population Dilemma

The most important factor in the decline of environmental conditions within the coastal zone has been the unprecedented increase in human population growth, particularly in the southeastern United States.

Presently more than half of the United States' population (153 million people) lives in coastal communities adjacent to the more than 66,645 miles of estuarine and coastal shoreline (Crossett et al. 2004). The U.S. population has increased by 33 million (28 percent) since 1980 and is expected to increase by anther 12 million by 2015 (U.S. EPA 1999; Crossett et al. 2004). The greatest rate of population change has been in the southeastern United States (58 percent increase) followed by the Pacific (46 percent) and Gulf of Mexico (45 percent) coastal regions (Crossett et al. 2004). Increased coastal population is not only a problem in the United States, it is also a problem globally, as more than 55 percent of the world's population lives within 50 miles of the coast, 33 of the 50 largest cities in the world are located in coastal areas, and more than 80 percent of world commerce is transported by ships (Dean 1997). The compression of more than 50 percent of the population into the coastal zone, which represents only 8 percent of the planet's surface, creates a dilemma for environmental managers, who are faced with the daunting task of trying to maintain environmental quality in the wake of unbridled urbanization and population growth (Dean 1997). Society places enormous value on coastal areas for living, working, and recreating; coastal waters support 28.3 million jobs and generate $54 billion in goods and services annually (U.S. EPA 2000). Direct ocean employment in the U.S. economy was 2.3 million people in 2000, with $117 billion in economic output (NOAA 2005). For example, there are more than 3,800 private marinas and more than 4,500 private charter boats along the coast of the United States (NOAA 1990). Direct ocean and coastal related employment (e.g., employment directly dependent on the oceans) grew by 18.5 percent from 1.9 million jobs in 1990 to 2.3 million jobs in 2000 (Colgan 2004). Direct ocean- and coastal-related employment wages increased by 46.3 percent, from $38.1 billion in 1990 to greater than $55.7 billion in 2000. The coastal region wage increase of 46.3 percent from 1990–2000 lagged behind overall U.S. wage increases of 76.2 percent, reflecting the creation of low-wage paying service industry jobs associated with tourism and recreation sectors within the coastal zone. There are about 44,000 miles of outdoor public recreation areas along the Atlantic, Pacific, and Gulf of Mexico coasts. The importance of these areas for ecotourism and similar leisure activities is increasing, as reflected by an increase of 41.4 percent in the number of tourism and recreational jobs created in the coastal zone between 1990–2000 (Colgan 2004). Nontourism employment declined by 136,000 jobs between 1990–2000, reflecting reduced military facilities for ship building and navigation, increased worker productivity in maritime transportation and oil/gas exploration, and declines in the commercial U.S. fisheries industry (Colgan 2004). This rapid growth in tourism in the coastal zone is placing increased demands on coastal ecosystems as a result.

Coastal Zone Management: Where Political Realities Shape Public Policy

Coastal communities are faced with enormous constraints and critical choices as they deal with rampant population growth. The lure of increased tax revenue for coastal municipalities is balanced with the stark reality of increased infrastructure costs, as communities strain to deal with the increased demands for roads, schools, sewer systems, and water. These urban services and resources generate property tax growth borne largely by citizens. Politicians face the daunting task of attempting to control growth and alienating the powerful real estate and development lobbies, or promoting unregulated development, which inevitably will cause taxes to rise and likely create discontent among the voting public. The doubling of tax bills in less than 7–10 years that has occurred in certain coastal counties can be particularly hard on people with fixed incomes, such as retirees, who compose a significant portion of the in-migrant population of coastal communities. Elected officials who take a stand against uncontrolled growth may find themselves in conflict with real estate or community interests, or both. For example, in Mount Pleasant, South Carolina, Mayor Harry Hallman introduced legislation passed by the town council that has placed a 3 percent cap on new construction, this being the estimated sustainable rate of population growth in the town (based on projected school, road, water, and sewer demands) (Walker 2005). The cap was opposed by real estate interests but supported by environmental and citizen groups. Similar "battle lines" have been drawn in nearly every coastal community in the United States. Elected officials everywhere must seek to achieve (or restore) the delicate balance between the economic rewards perceived to accrue from growth and development on the one hand, and managing growth and development so as to sustain cultural and environmental values and public services, i.e., perceived as quality of life, on the other. Potential tax increases to support public services necessitated by new development is also often an issue.

Extreme views are reflected in numerous politically divisive issues that may galvanize public opinion one way or the other, and lead to demands for political action that usually satisfy particular interests within the community. Issues of equity balance individual property rights against the common good (Raymond 2003), leaving politicians searching for answers that are based on facts rather than conjectures, and objective analysis rather than bias and special interests. Many believe that science can provide the objective information needed to inform and create fair resolution of many coastal land use controversies. In turn, science can help shape perceived human values and opinions in defining human dimensions in land use decisions. Human dimensions are thus the integration and balance of our social, political, economic, and environmental systems at work every day in coastal communities (Kennedy and Thomas 1995).

The Need for Sound Environmental Science

At the heart of most coastal land use issues lies a critical divide between environmental science and policy. The public may debate land use issues, but ultimately society looks to science to provide a rational basis for decision-making and for determining the impacts—both positive and negative—of those decisions. In the end, all parties understand that land use decisions should protect the natural resources and fragile coastal ecosystems that have attracted so many people to the coast in the first place. The common ethic is that the coastal environment is important and worth protecting. "Sound science" must arise from the sea of diverse and often divisive public opinions, as a means of answering the question that ultimately will determine the environmental and cultural integrity of our nation's coast: How much development is enough? Science may not always directly answer this specific question but it certainly provides the context for the public to make an informed choice in setting limits on growth and development. Science can also play a role in identifying the likely outcomes of land use decisions and resulting consequences of society's decisions. In effect, science can influence policy growth limits but not define it per se (Sandifer and Rosenberg 2005).

Sound science must be manifest within monitoring and assessment programs throughout the United States. Recent assessments of coastal conditions suggest that human health risks from sewage, eutrophication, and toxic effects of persistent organic compounds and emerging "contaminants-of-concern," such as pharmaceuticals, personal care products, and new generation pesticides, are also increasing and should be cause for alarm (Eilperin 2005). Recognitions of such threats has helped lead to efforts to develop guidance for implementation of large-scale monitoring programs (Bricker et al. 1999; U.S. EPA 1991; 1997; 1999) that would provide the critical information for integrated assessments and evaluation of coastal ecosystem health at different scales (e.g., locally, regionally, and nationally) such as the National Coastal Condition Report (U.S. EPA, 2004). Most national and regional monitoring programs can determine the status of the coastal water, sediment and resource conditions, including identification of contaminants or stressors of concern such as nutrients, microbes, and chemical contaminants, but lack specificity in quantitative linkage to land use sources of these different types of pollution. Watershed-based studies that utilize Total Maximum Daily Loading (TMDL) modeling approaches such as source tracking tools and hydrological and hydrodynamic modeling can establish quantitative linkages back to land based pollution sources (Kelsey et al. 2004; Nelson et al. 2006). In turn, results from these bioassessment and TMDL programs can be used to redirect zoning efforts to minimize the impacts of nonpoint source (NPS) runoff from urban areas into fragile coastal ecosystems. This has led to development of best management practices (e.g., vegetative buffer zones, swales, and retention ponds) and zoning practices (e.g., lot size, building setbacks and storm water utilities) that are designed to be one-size-fits-all models for containing and managing NPS pollution emanating from urbanization of coastal habitats. This can result in better coordinated multi-agency monitoring efforts to effectively address future coastal development.

"Paving paradise and putting up a parking lot," i.e., dramatically increasing the amount of impervious surface on the landscape, is characteristic of coastal development and has caused enormous, often poorly controlled alterations to local hydrology. This, in turn, has been the principal factor forcing increased loading of nutrients, microbes, and chemical contaminants from the land to coastal receiving waters and associated ecosystems (Holland et al. 2004; Kelsey et al. 2004).

More than simply an aesthetic resource to drive coastal development, coastal aquatic ecosystems are among the earth's most important sources of energy, biological activity, water, diversity, and biomass production. They supply food, oxygen, and other natural products critical for human existence, and interactions between the oceans and atmosphere shape the climate and weather, as well as commerce. Today, we recognize the coastal zone not only for these attributes, but also for its almost infinite diversity of life forms and processes—and the extraordinary potential for these natural resources to be harnessed for human welfare (Sandifer et al. 2004).

We are just beginning to understand the numerous and complex ways in which humans can affect the oceans, and the oceans, in turn, can affect human and environmental health in coastal ecosystems. The United States and its coastal resources annually contribute over a hundred billion dollars to the economy. A myriad of ecological services, worth billions of dollars, are provided free of charge by our nation's coastal ecosystems. Although the U.S. coastal zone encompasses only 25 percent of the nation's land area, more than 50 percent of the population lives and works here. In addition, ocean-based tourism is among the fastest-growing components of the U.S. coastal economy (Colgan 2004). Not surprisingly, coastal population densities are several times higher than in the rest of the nation, and coastal urban sprawl is consuming land at three or more times the rate of population growth. These trends are projected to continue and may accelerate, permanently altering large portions of the coastal landscape and potentially impacting marine ecosystems and public health in irreparable ways. In addition, coastal heritage and culture are rapidly being replaced by homogenous suburbs, golf courses, and resorts, devoid of any sort of regional context. Museums cannot replicate the traditional cultures and communities being lost to development. Nor can they adequately depict the lives of the people who lived in them. These and other aspects of human dimensions can be readily lost in a multi-faceted debate about coastal development. Human dimensions are shaped by education and knowledge of the value of coastal ecosystems and the related cultural heritage that evolves as a result (Bright and Tarrant 2002). For human dimensions to have merit, we as a society must understand, value, and preserve those facets of greatest value to society.

Estuaries: Where Urban Land Use Meets the Sea

Estuaries—those places where freshwater rivers meet and mix with the salt water of the ocean—are dynamic environments renowned for their ecological complexity, biological productivity, and seafood harvests, as well as the critical nursery habitat they provide to numerous ecologically and economically important species. Linking the land to the sea, the shallow tidal creeks and embayments along the shores of larger estuaries are the first zone impacted by chemical and microbial pollutants washed or released into estuaries (Fulton et al. 1996; Sanders et al. 2002; Holland et al. 2004).

Estuarine and coastal processes increasingly are being affected by urban development, with subsequent impacts on coastal ecosystems and the humans who live, work, and recreate there. Principal sources of pollution are urban runoff, municipal sewage discharges, agricultural runoff, atmospheric deposition of airborne pollutants, and industrial wastewater. Other causes of degradation include shoreline modification, dredging, overfishing, ballast water discharges, introduction of invasive species, high-density recreational use, and deposition of wind-carried, atmospheric pollutants. Increasing incidences of beach closures, fish and shellfish consumption advisories, harmful algal blooms, and occurrence of toxic chemicals and pathogenic microorganisms in coastal waters, sediments, and biota are indicative of the extent of the problem. Stress-induced changes in marine ecosystems allow microbial pathogens transmitted via water, food, or other vectors to be harbored in animal reservoirs and to threaten human health.

Throughout the coastal United States, estuarine organisms are being exposed to multiple stressors, including increased loading of chemical contaminants, nutrients, and bacteria and viruses. In addition, the rapid pace of landscape modification in coastal areas results in developed coastal watersheds with increased impervious surfaces, which can alter hydrography, hydrological budgets, and water delivery. Parking lots, roads, and rooftops are the major types of impervious surface in coastal watersheds (Arnold and Gibbons 1996; Sanger et al. 1999a, b; Holland et al. 2004). Changes in the amount of impervious surface in a watershed may increase the volumes of runoff (McCall et al. 1987), alter dissolved and particulate contaminant concentrations (McCall et al. 1987), and increase the delivery rate of runoff into coastal receiving waters. When the amount of impervious surface in a watershed reaches 20–35 percent of total watershed area, the loading of chemical and other contaminants is usually sufficient to alter water quality and to impact indigenous fauna and flora (Schuler et al. 1994; Sanger et al. 1999a, b; Holland et al. 2004). Ultimately, the frequency of land-based discharges from upland areas into coastal watersheds increases, which may, in turn, increase the diversion of water inflow to groundwater supplies. Recent studies have demonstrated that groundwater discharges (greater than 40 percent) may account for a significant portion of the freshwater flow into coastal watersheds of the south Atlantic (Moore 1996). Alterations in hydrological budgets may result not only in dwindling subsurface water supplies for drinking water, but also in increased drought or flood potential for coastal areas (USGS 2004).

Paradise Lost

Throughout the United States, there are reminders of the importance of what happens when we fail to control coastal development and growth. Poorly planned, massive coastal development ultimately results in a coastal zone where marine organisms are faced not only with the normal rigors of a constantly fluctuating, physicochemical water-quality environment (e.g., salinity, dissolved oxygen, pH, and temperature), but also more pronounced fluctuations combined with cumulative anthropogenic stressors (e.g., contaminants, nutrients, and microbiological agents). In many instances the level of any single contaminant or stressor may not by itself pose significant effects, but when taken together, direct adverse impacts as well as a general decrease in ecological fitness may result (De Lorenzo et al. 1999; Pennington and Scott: 2001; Christyl et. al. 2004). Ultimately, coastal habitats must cope with an increase in the number and types of harmful algal blooms, increases in the area of water and sediment quality impairment, and increased incidence of fish disease leading to declines in the stocks of fish and shellfish (e.g., lobster, crabs and corals). The National Marine Fisheries Service estimates that many U.S. fish stocks are overused or below levels needed to sustain long-term potential yields (NMFS 1992; Sissenwine and Rosenberg 1993). A review of seven coastal regional reports in the United States showed that only 45 percent of the 73 federally managed fisheries were at sustainable capacities (Ward et al. 2001).

Areas such as the Chesapeake Bay have been overdeveloped to the point that the oyster industry is on the verge of collapse, and the filtering capacity of the bay provided by oysters has declined dramatically. As a result, turbidity has increased, reducing the submerged aquatic vegetation growth that is vital for the health of the living marine resources of the bay. Eutrophication from overdevelopment of adjoining uplands has been the most important factor in the decline of the Chesapeake Bay ecosystem; the price tag for restoration has been placed at more than $18 billion, with the cost for the state of Maryland exceeding $7 billion (MD NRD 2001). Similarly, in South Florida, intensive agricultural and suburban development in the coastal zone has resulted in eutrophication and chemical runoff. South Florida provides a model of impacts from rapid coastal urbanization. More than 3,000 people/square mile reside in the region, and more than 14,000 tons of pesticides are applied to the South Florida landscape annually (Miles and Pfeuffer 1997). Insecticide use (38 percent of the total) in South Florida is nearly double the national average. Management of dwindling water supplies has become the responsibility of the Water Management Districts (which were originally created for flood control). Because of the massive redistribution of water flows, coupled with attendant nutrient and chemical contaminant issues within the region (LaPoint et al. 1998; Scott et al. 2002; Fulton et al. 2004), costly long-term restoration efforts are now necessary for the Everglades and Florida Bay. These examples underscore the need for "sound science" that will help us maintain coastal environmental integrity and sustainable coastal development. Sound science has been used to guide restoration efforts in South Florida and the Chesapeake Bay. The price tag for coastal restoration far excedes the cost of preventive management of coastal environmental health. The lesson learned in both South Florida and the Chesapeake Bay is that sound science should also be used in a predictive or forecast fashion as development occurs to develop land use planning and zoning decisions that protect the environment and prevent coastal ecosystem degradation.

Paradise Gained: The Need for Action

In human health care, the cost of restoring health may sometimes exceed the cost of preventive care and health maintenance by a factor of more than 1000:1. Let us as a society take measure of the value of a preventive health care approach in managing coastal environments in the future (Schaeffer 1996; Rapport et al. 1998). Ecosystems are fragile, and once impacted, they may be slow to recover (Schaeffer 1996). Hence, we ought to apply the "precautionary principle" when dealing with the uncertainties of the science needed to make effective land use decisions, because the volume of information needed to make science-based decisions may exceed our capacity to deliver the "sound science" within the time frame required for the decision to be effective as a policy or action. Sandifer and Rosenberg (2005) note that the differing degrees of uncertainties associated with scientific findings and the strict application of the "precautionary principle" may lead to gridlock in resource use and management and conservation issues, and they rather prefer a more common sense "precautionary approach" that focuses on (1) application of the best available information and management practices from the inception of an issue, (2) weighing and balancing the scientific uncertainty versus the likelihood of risk of harm to resources or ecosystems, and (3) using an adaptive environmental management approach to continually gather scientific data for periodic reassessments and modifications of management practices. This "precautionary approach" can lead to the development of generic land use management models for a number of common land use activities such as residential and urban uses, golf courses, and agricultural NPS runoff controls. It is important that science-based land use models have a certain generic quality, allowing them to be translatable from one region to another and from one estuary to the next. It is also critical that these science-based approaches incorporate human dimensions in consideration of identified environmental impacts from development, because it is what society wants and expects out of the natural resource utilization within the coastal zone that ultimately defines land use policies and decisions.

Several studies, such as the Urbanization in Southeast Estuarine Systems (USES) (Vernberg et al. 1999, 2001), the Tidal Creek Project (Holland et al. 2004) and the Land Use-Coastal Ecosystems Study (LU-CES) have focused on the mission of providing sound science to guide coastal land use decision-making. All of these studies have utilized ecosystem level spatial and temporal scale analysis to produce data that leads to understanding and models of ecosystem structure and function to elucidate cause and effect relationships between land use and coastal condition. All of these studies have focused on defining clear linkages between land use and environmental health within estuarine systems, beginning at the tidal creek scale (e.g., the Tidal Creek Project) and broadening to include both the high-salinity (USES) and riverine (LU-CES) estuaries with tidal ranges from 2 to 3 meters during spring tide. The chapters in this book have chronicled the state of our knowledge on the impacts of development on our fragile coastal ecosystems. They have, in several cases, directly or implicitly, suggested how best to manage coastal ecosystems under developmental pressure and how to address, through sound science, some of the present and future environmental issues associated with urbanization in the coastal zone. What is clear is that we must apply this knowledge soon, even in the face of uncertainty. The current pace of coastal development does not permit us the luxury of delaying (Brown 1996).

Acknowledgments

The authors wish to thank Dr. Malcolm Meaburn, Dr. Mike Fulton, and Paul Comar of NOAA/NCCOS Center for Coastal Environmental Health and Biomolecular Research, and Dr. Gary Kleppel of the State University of New York at Albany for their helpful suggestions and extensive review of the manuscript.

References

Arnold, C. L. Jr. and J. Gibbons. 1996. Impervious surface coverage: the emergence of a key environmental indicator. *Journal of the American Planning Association* 62:243–258.

Bricker, S. B., C. G. Clement, D. E. Pirhalla, S. P. Orlando, D. R. G. Farrow. 1999. National estuarine eutrophication assessment: Effects of nutrient enrichment in the nation's estuaries. National Ocean Service, NOAA, Silver Spring, MD. 71 p.

Bright, A. D., and M. A. Tarrant. 2002. Effect of environment-based coursework on the nature of attitudes toward the Endangered Species Act. *Journal of Environmental Education* 33:10–19.

Brown, L. R. 1996. The acceleration of history, pp. 3–20. In, Brown, L. R. et al. (eds.), The State of the World 1996: A Worldwatch Institute Report on Progress Toward a Sustainable Society. Worldwatch Institute, Washington, DC; W.W. Norton, New York, NY.

Christyl, T. J., P. Pennington, M. DeLorenzo, K. J. Karnaky, and G. I. Scott 2004. Effects of multiple atrazine exposure profiles on hemocyte DNA integrity in the eastern oyster. *Bulletin of Environmental Contamination and Toxicology*, 73:404–410.

Colgan, C. S. 2004. Employment and wages for the U.S. ocean and coastal economy. *Monthly Labor Review*, November 2004:24–30.

Crossett, K. M., T. J. Culliton, P. C. Wiley and T. R. Goodspeed. 2004. Population trends along the coastal United States: 1980–2008. NOAA Technical Memo, National Oceanic and Atmospheric Administration, National Ocean Service, Special Project Office, Silver Spring, MD. 54 p.

Dean, J. M. 1997. A crisis and opportunity in coastal oceans: coastal fisheries as a case study, pp. 81–88. In, Vernberg, F. J, W. B. Vernberg, and T. Siewicki (eds.), Sustainable Development in the Southeastern Coastal Zone. Belle W. Baruch Library in Marine Science 20.

DeLorenzo, M. E., J. Lauth, P. L. Pennington, G. I. Scott, and P. E. Ross. 1999. Atrazine effects on the microbial food web in tidal creek mesocosms. *Aquatic Toxicology* 46:241–251.

Eilperin, J. 2005. Pharmaceuticals in waterways raise concern: Effects on wildlife, humans questioned. *Washington Post*, June 23, 2005, p. A03

Fulton, M. H., G. T. Chandler, and G. I. Scott. 1996. Urbanization effects on the fauna of a southeastern U.S.A. estuary bar-built estuary, pp. 477–504. In, Vernberg, F. J, W. B. Vernberg, and T. Siewicki (eds.), Sustainable Development in the Southeastern Coastal Zone. Belle W. Baruch Library in Marine Science 20.

Fulton, M. H., G. I. Scott, M. E. DeLorenzo, P. B. Key, D. W. Bearden, E. D. Stroizier, and C. J. Madden. 2004. Surface water pesticide movement from the Dade County Agricultural area to the Everglades and Florida Bay via the C-111 Canal. *Bulletin of Environmental Contamination and Toxicology* 73:527–34.

Holland, A. F., D. M. Sanger, C. R. Gawle, S. B. Lurberg, M. S. Santiago, G. H. M. Riekerk, L. E. Zimmerman, and G. I. Scott. 2004. Linkages between tidal creek ecosystems and the landscape and demographic attributes of their watersheds. *Journal of Experimental Marine Biology and Ecology* 298:151–178.

Kelsey, H. E., G. Scott, D. E. Porter, B. Thompson, and L. Webster. 2003. Using multiple antibiotic resistance and land use characteristics to determine sources of fecal coliform bacterial pollution. *International Journal of Environmental Monitoring and Assessment* 81:337–348.

Kennedy, J. J. and J. W. Thomas. 1995. Managing natural resources as social values, pp. 311–321. In, Knight, R. L. and S. F. Bates (eds.), A New Century for natural Resource Management. Island Press, Washington, DC.

LaPoint, T. W., J. H. Rodgers Jr., J. J. Delphino, T. D. Atkeson, and S. C. McCutcheon. 1998. Advisory panel report on the workshop: Ecological Risk of Toxic Substances in South Florida Ecosystems, Roz and Cal Kovens Conference Center, October 20–21, Florida International University, Boca Raton, FL, 45 p.

Maryland Department of Natural Resources (MD NRD). 2001. New fiscal analysis puts Maryland Bay restoration costs at $7 billion. Maryland Department of Natural Resources, http://dnrweb.dnr. State.md.us/bay/res_protect/c2k/costs.aps., 2 pp.

McCall, E. C. Jr., G. I. Scott, and J. M. Hurley. 1987. A comparison of agricultural nonpoint source runoff from black plastic and conventional tillage tomato test plots. Technical Report # G1251–07, U.S. Geological Survey, U.S. Department of the Interior, Reston, VA. 51 p.

Miles, C. J. and R. J. Pfueffer. 1997. Pesticides in canals of South Florida. *Archives of Environmental Contamination and Toxicology* 32:337–345.

Moore, W. S. 1996. Large groundwater inputs into coastal waters revealed by ^{226}RA enrichments *Nature* 380:612–614

National Marine Fisheries Service. 1992. Our living oceans: Report on the status of U.S. living marine resources. NOAA Technical Memo NMFS–F/SPO–2, National Oceanographic and Atmospheric Administration, Silver Spring, MD. 148 p.

Nelson, K. A., G. I. Scott, and P. F. Rust. 2006. A multivariate approach for evaluating major impacts on water quality in Murrells Inlet and North Inlet, South Carolina. *Journal of Shellfish Research,* 24:1241–1251.

NOAA. 1990. Estuaries of the United States vital statistics of a natural resource base. National Oceanic and Atmospheric Administration. Rockville, MD. 79 p.

NOAA. 1999. Trends in U.S. coastal regions, 1970–1998. NOAA, NOS, Special Projects Office, Silver Springs, MD. 29 p.

NOAA. 2005. Economic statistics for NOAA, 4th Edition. U.S. Dept of Commerce, NOAA. Silver Spring, MD. 4 p.

Pennington, P. L. and G. I. Scott. 2001. The toxicity of atrazine to the estuarine phytoplankter Pavlova sp. (prymnesiophyceae): Increased sensitivity after chronic exposure. Environmental Toxicology and Chemistry 20(10):2237–2242

Rapport, D., R. Constanza, P. R. Epstein, C. Gaudet, and R. Levins. 1998. Ecosystem Health. Blackwell Science, Oxford, UK. 372 p.

Raymond, L. 2003. Private Rights in Public Resources: Equity and Property Allocation in Market-Based Environmental Policy. Resources for the Future Press, Washington, DC. 252p.

Sanders, M., S. Siversten, and G. Scott. 2002. Origin and distribution of polycyclic aromatic hydrocarbons in surface sediments from the Savannah River. *Archives of Environmental Contamination and Toxicology* 43:438–448.

Sandifer, P. A. and A. A. Rosenberg. 2005. Practical recommendations for improving the use of science in marine fisheries management, pp. 197–210. In, Witherell, D. (ed.), Managing our Nations Fisheries II—Focus on the Future. Proceedings of a conference of fisheries management in the United States, Washington, DC, March 24–26, 2005. 283 p.

Sandifer, P. A., A. F. Holland, T. K. Rowles, and G. I. Scott. 2004. The oceans and human health. Guest editorial. *Environmental Health Perspectives* 112 (8):454–455.

Sanger, D. M., A. F. Holland, G. I. Scott. 1999a. Tidal creek marsh sediments in South Carolina estuaries I. Distribution of trace metal contaminants. *Archives of Environmental Contamination and Toxicology* 37:445–457.

Sanger, D. M., A. F. Holland, and G. I. Scott. 1999b. Tidal creek marsh sediments in South Carolina estuaries II. Distribution of organic contaminants. *Archives of Environmental Contamination and Toxicology* 37:458–471.

Schaeffer, D. J. 1996. Diagnosing ecosystem health. *Ecotoxicology and Environmental Safety* 34:18–34.

Schueler, T. R. 1994. The importance of imperviousness. *Watershed Protection Techniques* 1(3):37–48.

Scott, G. I., M. H. Fulton, E. F. Wirth, G. T. Chandler, P. B. Key, J.W.Daugomah,D.Bearden,K.W.,Chung,E.D.Strozier,M.E.DeLoranzo, S. Sivertsen, A. Dias, M. Sanders, J. M. Macauley, L. R. Goodman, M. W. LaCroix, G. W. Thayer, and J. Kucklick. 2002. Toxicological studies in tropical ecosystems: An ecotoxicological risk assessment of pesticide runoff in south Florida estuarine ecosystems. *Journal of Agricultural Food Chemistry* 50(15):4400–4408.

Sissenwine, M. P. and A. A. Rosenberg. 1993. Marine fisheries at a critical juncture. *Fisheries* 18(10):6–14.

U.S. Environmental Protection Agency (U.S. EPA). 1991. Portraits of our coastal waters—Supplement to National Water Quality Inventory. Report from the EPA Regions EPA 503/2–91–004. U.S. Environmental Protection Agency, Office of Water, Washington, DC. 31 p.

U.S. Environmental Protection Agency (U.S. EPA). 1997. Estuarine and Coastal Marine Waters Bioassessment and Biocriteria Technical Guidance. U.S. Environmental Protection Agency, Office of Science and Technology, Washington, DC. 300 p.

U.S. Environmental Protection Agency (U.S. EPA). 1999. Clean Water Action Plan: Coastal Research and Monitoring Strategy. U.S. EPA,

NOAA, DOI, and Department of Agriculture; EPA Office of Water; Washington, DC. 70 p.

U.S. Environmental Protection Agency (U.S. EPA). 2000. Liquid Assets 2000: America's Water Resources at a Turning Point. EPA–840–B–00–001. U.S. Environmental Protection Agency, Office of Water, Washington, DC. 20 p.

U.S. Environmental Protection Agency (U.S. EPA). 2004. National Coastal Condition Report II. EPA-620/R-03/002. U.S. Environmental Protection Agency, Office of Research Development and Office of Water, Washington, DC: 286 p.

U.S. Geological Survey (USGS). 2004. South Florida restoration—hydrology—Where did the water go before it was managed? USGS Web site, http://ofia.usgs.gov/sfrf/, 4 pp.

Vernberg, F. J., W. B. Vernberg, D. E. Porter, G. T. Chandler, H. N. Mckellar, D. Tufford, T. Siewicki, M. Fulton, G. Scott, D. Bushek, and M. Wahl. 1999. Impact of coastal development on land-coastal waters, pp. 612–622. In, Ozhan, E. (ed.), Land Ocean Interactions: Managing Coastal Ecosystems, MEDCOAST, Middle East Technical University, Ankara, Turkey.

Vernberg, F. J. and W. B. Vernberg. 2001. The Coastal Zone: Past, Present and Future. University of South Carolina Press, Columbia. 191 p.

Walker, T. 2005. Suit fails to topple Mt. Pleasant building cap. *Charleston Post and Courier*, July 9, 2005, section B, p. 1.

Ward, J. M., T. Brainerd, and M. Milazzo. 2001. Identifying Harvest Capacity and Over Capacity in Federally Managed Fisheries: A preliminary Qualitaitve Report. Office of Science and Technology and Office of Sustainable Fisheries, Department of Commerce, NOAA, NOAA Fisheries, Silver Spring, MD.

Index